FOUNDATION CHEMISTRY

Exam-style questions with answers

Atomic Structure, Amount of Substance, Bonding, Periodicity, Introduction to Organic Chemistry, and Alkanes.

Dr. Muhammad Eesa

Table of Contents

Introduction

Many students of chemistry are faced with a serious problem: while there are very good textbooks that help the student learn the subject, there are no published books that help the student learn and practise solving problems.

This book is an attempt to remedy the situation. It provides a large number of problems especially formulated to familiarise the student with exam-style questions. The book can be useful to any student studying foundation chemistry, whether at high school or as part of their university degree. Although the book is based on the AS level specifications (in the UK), it can also be used by any student studying chemistry at a foundation level.

The topics covered in this book are: atomic structure, amount of substance, bonding, periodicity, introduction to organic chemistry, and alkanes. Each topic is dealt with using exam-style questions. Each question is followed by a full answer or solution. In most cases, however, it may be useful to give the student a more detailed explanation of the answer. In such cases, the answer is followed by a 'background' section which serves to explain in detail the points covered by the answer and any relevant information deemed relevant or useful.

Although every attempt has been made to ensure that the questions cover all the areas within each topic (at a foundation level), this book should be regarded as a support material and not as a substitute for the student's textbook or class notes.

This book can also be used to test your knowledge of the topics covered here. When you read a question, try to answer it in your own way, and then compare your answer to the answer given in the book. If any part of your answer is incorrect, you should identify the mistake and learn how to avoid it.

I hope this book will improve the reader's understanding of the topics covered and help chemistry students achieve high grades in their exams.

The author

1. ATOMIC STRUCTURE

Question 1.1

Give the relative charge and relative mass of each of the sub-atomic particles.

Answer

Sub-atomic particle	Relative charge	Relative mass
Proton	+1	1
Neutron	0	1
Electron	-1	1/1840

Background

The three subatomic particles that make up an atom are the proton, the neutron and the electron. The charge and the mass of a subatomic particle are measured relative to the charge and the mass of one proton. The mass of the neutron is nearly equal to the mass of the proton, while the mass of the electron is negligible. The proton carries a positive electric charge, the electron a negative charge, and the neutron is neutral. The nucleus of the atom contains protons and neutrons; therefore, the nucleus carries a net positive charge.

Question 1.2

State the number of protons, neutrons and electrons in the following atoms:

$^{12}C, ^{14}C, ^{35}Cl, ^{37}Cl$

Answer

Atom	Number of protons	Number of neutrons	Number of electrons
^{12}C	6	6	6
^{14}C	6	8	6
^{35}Cl	17	18	17
^{37}Cl	17	20	17

Background

Carbon is the 6th element in the periodic table; it has an atomic number of 6, which means that the number of protons in a carbon nucleus is 6. Since the number of electrons in any neutral atom is

2

equal to the number of protons, the carbon atom has 6 electrons. The number of neutrons can be calculated as the difference between the mass number and the atomic number of the element. For ^{12}C, the atomic number is 6 and the mass number is 12; therefore, the number of neutrons is 12 − 6 = 6. Similarly, for ^{14}C the number of neutrons is 14 − 6 = 8.

Chlorine has an atomic number of 17, which means that a chlorine atom has 17 protons and 17 electrons. The number of neutrons in a ^{35}Cl atom is 35 − 17 = 18. The number of neutrons in a ^{37}Cl atom is 37 − 17 = 20.

Question 1.3

(i) State the meaning of the term *isotopes*.

(ii) Explain why isotopes of the same element have the same chemical properties.

Answer

(i) Isotopes are atoms with the same number of protons but different numbers of neutrons.

(ii) Isotopes of the same element have the same number of protons and the same number of electrons. Since the chemical properties of an atom are determined by the number of electrons it contains, isotopes of the same element react chemically in exactly the same way.

Background

An element is identified by its atomic number (the number of protons). Atoms that have the same number of protons but different numbers of neutrons are called isotopes. Since isotopes are atoms of the same element, we use the same chemical symbols to represent them, while also indicating their mass numbers. For example, chlorine has two isotopes, ^{35}Cl and ^{37}Cl, with mass numbers of 35 and 37, respectively.

The chemical properties of an element are determined by the number of electrons in its atom; therefore, the isotopes of an element have the same chemical properties since they have the same number of electrons.

Question 1.4

An atom X contains half as many protons as a selenium (Se) atom and twice as many neutrons as a fluorine ^{19}F atom. Deduce the identity of X, including its mass number and atomic number.

3

Answer

The atom is the chlorine isotope $^{37}_{17}Cl$.

Background

From the periodic table we can see that the number of protons in a selenium atom is 34 (atomic number 34). Therefore, the number of protons in X is 34/2 = 17. The element with atomic number 17 is chlorine. But chlorine has two isotopes with different numbers of neutrons, so to determine which chlorine isotope the atom X is, we have to find the number of neutrons. Fluorine ^{19}F has a mass number of 19, which means it has a total of 19 protons and neutrons; it also has an atomic number of 9, which means it has 9 protons. Therefore, the number of neutrons in ^{19}F is 19 − 9 = 10. X has twice as many neutrons as ^{19}F, i.e. 20. This means that the mass number of X is 17+20 (protons + neutrons), which gives 37. So, X is chlorine with an atomic number of 17 and mass number of 37, i.e. $^{37}_{17}Cl$.

Question 1.5

An atom has one more proton than, but the same number of neutrons as, an atom of ^{12}B. Deduce the symbol of the atom, including its mass number and atomic number.

Answer

The atom is the carbon isotope $^{13}_{6}C$.

Background

The atom of $^{12}_{5}B$ has 5 protons and 7 neutrons (12 − 5 = 7). This means that the atom we are looking for has 6 protons and 7 neutrons, i.e. the atomic number is 6 (which is carbon) and the mass number is 6 + 7 = 13. This gives $^{13}_{6}C$.

Question 1.6

In a mass spectrometer, the vaporised sample is ionised using an electron gun. What is the purpose of the ionisation step?

Answer

The gaseous atoms are ionised so that they can be accelerated, deflected, and detected.

Background

The mass spectrometer is used to measure the relative masses of the isotopes of an element and to find their relative abundances. The vaporised sample of the element is first ionised by bombarding the atoms with high energy electrons, which in turn knock out electrons from the atoms turning them into positive ions. The positive ions are then attracted towards negatively charged plates and are thus accelerated to a high speed as they move towards the magnetic field. The magnetic field then causes the beam of charged ions to be deflected into an arc of a circle. The amount of deflection of an ion depends on the ratio of its mass, m, to its charge, z. A magnetic field cannot deflect neutral atoms – only charged ions can be deflected. As a result, the atoms must be ionised before they can be deflected by the magnetic field. The deflected ions are then detected using an ion-current detector. As the ions strike the detector, they create an electric current. This enables the detection of the ions and the measurement of their abundance.

Question 1.7

Mass spectrometry is used to measure the relative mass and abundance of the isotopes of an element. Explain how the abundance is measured in a mass spectrometer.

Answer

The abundance of an isotope is measured using an ion-current detector. When the ions of an isotope strike the detector, they create a current which is proportional to the abundance of the ions.

Background

The amount of current created in the detector by the stream of ions depends on the amount of ions reaching the detector. By measuring the electric current generated, we can therefore measure the abundance of the ions causing the current.

Question 1.8

Explain how ions with different m/z values are directed onto the detector in a mass spectrometer.

Answer

By varying the strength of the magnetic field, ions with different m/z values are deflected by sufficient amounts to reach the detector.

Background

In a mass spectrometer, the degree of deflection caused by the magnetic field depends on the

- Mass of ion (m)
- Charge of ion (z)
- Strength of magnetic field

Assuming the ions have the same charge (e.g. 1^+), the heavier the ions the stronger the magnetic field that is required to cause sufficient deflection in their movement so that they can reach the detector. Therefore, lighter isotopes require weaker magnetic fields, and heavier isotopes require stronger magnetic fields.

Question 1.9

Explain how the m/z ratio of a particular ion is measured in the mass spectrometer.

Answer

The mass spectrometer measures the m/z of an ion from the strength of the magnetic field at which the ion is detected.

Background

As explained in Question 1.8, there is a relationship between the m/z value of an ion and the strength of the magnetic field required to cause enough deflection so that the ion can be detected.

Question 1.10

The mass spectrum of zirconium is shown in Figure 1.1. Use the information in the figure to calculate the relative atomic mass of zirconium.

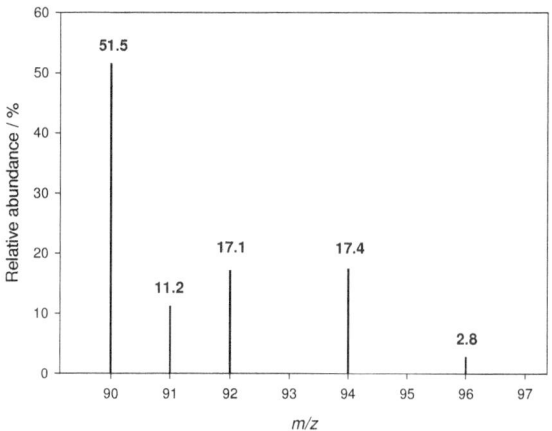

Figure 1.1: The mass spectrum of zirconium.

Answer

$$A_r \text{ of } Zr = \frac{(90 \times 51.5) + (91 \times 11.2) + (92 \times 17.1) + (94 \times 17.4) + (96 \times 2.8)}{100}$$

$$= 91.3$$

Background

The relative atomic mass of an element is calculated as the average of the relative atomic masses of all of its isotopes. The relative atomic masses of the isotopes of an element are obtained from the mass spectrum of the element. The mass spectrometer measures the ratio m/z for each isotope, where m is the relative atomic mass and z is the charge on the ion. For example, boron B has two isotopes, ^{10}B and ^{11}B. When a sample of boron is ionised, two positive ions are formed: $^{10}B^+$ and $^{11}B^+$ resulting in two peaks at m/z values of

$$\frac{10}{+1} = 10, \quad \text{and} \quad \frac{11}{+1} = 11$$

The mass spectrometer also measures the percentage (abundance) of each of the isotopes in the sample. The average relative atomic mass is then calculated from

$$A_r \text{ of an atom} = \frac{\Sigma(\text{mass of each isotope} \times \text{its percentage})}{100}$$

where Σ means sum.

7

Although most atoms lose only one electron in the mass spectrometer forming ions with a 1^+ charge, a small number of atoms can lose *two* electrons, thus forming doubly charged ions with a 2^+ charge. For example, a sample containing sulphur, ^{32}S, may produce a peak at $m/z = 16$, although the mass number of S is 32. This peak is caused by the ion $^{32}S^{2+}$ which has a relative atomic mass of 32 and a charge of 2^+, so

$$\frac{m}{z} = \frac{32}{+2} = 16$$

This explains why the ions $^{32}S^{2+}$ and $^{16}O^+$ both produce the same peak at $m/z = 16$ and, therefore, cannot be distinguished in a mass spectrometer.

Question 1.11

The mass and relative abundance of four isotopes of an element X are given below.

Relative abundance	1.5	2.5	3.0	4.5
m/z	188	189	190	192

Use this data to calculate the relative atomic mass of X. Deduce the identity of X.

Answer

$$A_r = \frac{(188 \times 1.5) + (189 \times 2.5) + (190 \times 3.0) + (192 \times 4.5)}{1.5 + 2.5 + 3.0 + 4.5}$$
$$= 190.3$$

X is osmium Os.

Background

In this question, the relative abundance is used to calculate the relative atomic mass of X from the formula

$$A_r \text{ of an atom} = \frac{\Sigma(\text{mass of each isotope} \times \text{its relative abundance})}{\text{Sum of relative abundances}}$$

Using the periodic table, we find that the nearest relative atomic mass to the answer, 190.3, is that of osmium, Os = 190.2. The reason that our answer does not agree exactly with the actual A_r of osmium is that only four isotopes of Os have been used in the calculation above, while osmium has seven naturally occurring isotopes. However, the remaining three isotopes which have not been included in the table above are very rare.

Question 1.12

Chlorine has two isotopes, ^{35}Cl and ^{37}Cl. The mass spectrum of a sample of chlorine produces peaks with the following *m/z* values: 35, 37, 70, 72, and 74. Identify the ion responsible for each peak.

Answer

The peak at 35 is caused by $^{35}Cl^+$

The peak at 37 is caused by $^{37}Cl^+$

The peak at 70 is caused by $^{35}Cl-^{35}Cl^+$

The peak at 72 is caused by $^{35}Cl-^{37}Cl^+$

The peak at 74 is caused by $^{37}Cl-^{37}Cl^+$

Background

If a sample of a diatomic element, such as H_2, O_2, Cl_2, N_2, is analysed in a mass spectrometer, the sample will produce ions of the individual atoms as well as ions of the diatomic molecule. For example, a sample of chlorine will contain the ions of chlorine atoms (Cl) and small amounts of the ions of chlorine molecules (Cl_2). Since chlorine has two isotopes (^{35}Cl and ^{37}Cl), the two atoms in an ionised chlorine molecule can form any one of the following combinations:

$^{35}Cl-^{35}Cl^+$ (total mass = 35+35 = 70)

$^{35}Cl-^{37}Cl^+$ (total mass = 35+37 = 72)

$^{37}Cl-^{37}Cl^+$ (total mass = 37+37 = 74)

Therefore, three peaks corresponding to these three combinations will be produced in the mass spectrum of chlorine.

Note that the mass spectrum of chlorine may also show two very small peaks at *m/z* = 17.5 and 18.5 caused by the doubly charged ions $^{35}Cl^{2+}$ and $^{37}Cl^{2+}$ respectively.

Question 1.13

State one use of radioactive isotopes.

Answer

Carbon dating.

Background

Some isotopes have unstable nuclei. The unstable nucleus of the atom breaks down releasing energetic rays; this process is called *decay*. Such isotopes are called *radioactive isotopes*. Different radioactive isotopes decay at different rates. The rate of radioactive decay of an isotope is measured by the time taken for half of its radioactivity to decay. This time is called the *half-life* of the isotope. For example, carbon has a radioactive isotope ^{14}C with a half-life of 5730 years (i.e. it takes 5730 years for half the amount of ^{14}C in any sample of carbon to decay). In living organisms, there is a certain level of radioactivity due to ^{14}C. However, when the organism dies, the level of radioactivity starts to fall (since the decaying ^{14}C is not being replaced by fresh carbon containing ^{14}C). By measuring the present level of radioactivity in a dead organism, we can calculate the length of time for which the ^{14}C has been decaying and therefore determine the age of the organism. This technique is called *carbon dating*.

Question 1.14

Explain why high resolution mass spectrometers can be used to identify elements.

Answer

Because high resolution mass spectrometers can measure the relative atomic mass to several decimal places.

Background

Since atoms, apart from carbon-12, have relative atomic masses which are not whole numbers, any atom can be identified by its exact atomic mass measured to several decimal places. High resolution mass spectrometers can measure the masses of atoms to several decimal places, thus allowing us to identify elements by their exact relative masses.

Question 1.15

Mass spectrometry has been applied in space studies. State two such applications of mass spectrometry in space.

Answer

Mass spectrometry has been used to identify the elements in rock samples from Mars, and to analyse the composition of Halley's Comet.

Background

High resolution mass spectrometry is used to identify elements. Mass spectrometers carried on the Viking Martian lander have been used to identify the elements present in Martian rocks. Another space probe used mass spectrometers to identify the composition of Halley's Comet.

Question 1.16

Write the full electron arrangement for Ne, Mg, Mg^{2+}, Cl, Cl^-, and Ar.

Answer

Ne $1s^2, 2s^2, 2p^6$

Mg $1s^2, 2s^2, 2p^6, 3s^2$

Mg^{2+} $1s^2, 2s^2, 2p^6$

Cl $1s^2, 2s^2, 2p^6, 3s^2, 3p^5$

Cl^- $1s^2, 2s^2, 2p^6, 3s^2, 3p^6$

Ar $1s^2, 2s^2, 2p^6, 3s^2, 3p^6$

Background

In allocating electrons to atomic orbitals, lower energy levels are filled first; so the main energy level $n = 1$ is filled first, then $n = 2$, and so on. Each energy level contains sub-levels. The first level ($n = 1$) contains one sub-level (which is 1s), the second ($n = 2$) contains two sub-levels (2s, 2p), and the third ($n = 3$) contains three sub-levels (3s, 3p, 3d). Within each energy level, sub-levels of lower energy are filled first, so sub-level s (which holds up to 2 electrons) is filled first, followed by sub-level p (which holds up to 6 electrons), and then d (which holds up to 10 electrons).

When electrons are removed from an atom to form a positive ion, they are removed from the highest energy level first. So, to work out the electron arrangement of Mg^{2+}, remove the two outermost electrons in Mg. Note that the resulting electron arrangement of Mg^{2+} is identical to that of Ne, which is the nearest noble gas to Mg. Atoms usually lose, gain, or share electrons in order to attain the electronic structure of the nearest noble gas. Chlorine usually reacts by gaining an electron to form the negative chloride ion Cl^-, which has the same electron arrangement as argon Ar, the nearest noble gas to chlorine.

Question 1.17

Write the full electron arrangement for Ca and Ti.

Answer

Ca $1s^2, 2s^2, 2p^6, 3s^2, 3p^6, 4s^2$

Ti $1s^2, 2s^2, 2p^6, 3s^2, 3p^6, 3d^2, 4s^2$

Background

As explained in Question 1.16, lower energy levels are filled first, and within each main energy level, lower energy sub-levels are filled first. However, in titanium, Ti, the 4s sub-level is filled before the 3d sub-level. This is because, in a neutral atom, the energy of 4s is slightly lower than that of 3d, and so 4s is filled first. An element which has at least one ion with a part-filled d sub-level is called a *transition metal*. Ti is an example of a transition metal.

Question 1.18

The following diagram shows the successive ionisation energies of an element X.

(i)   Define the terms *first ionisation energy* and *second ionisation energy*.
(ii)   Explain how measuring the successive ionisation energies of an unknown element can be used to identify the element, and hence deduce the identity of X.
(iii)   Write an equation that represents the third ionisation of X.
(iv)   Explain why the ionisation energy is plotted using a logarithmic scale.

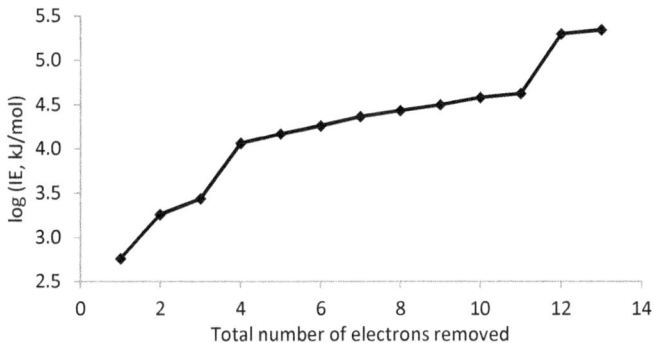

Figure 1.2: Successive ionisation energies of element X

Answer

(i) First ionisation energy is the energy required to remove one mole of electrons from one mole of gaseous atoms to form one mole of 1^+ ions. Second ionisation energy is the energy required to remove one mole of electrons from one mole of gaseous 1^+ ions to form one mole of gaseous 2^+ ions.

(ii) The electrons in the outermost shell are the easiest to remove as they are the furthest from the nucleus. The electrons in the next (inner) shell require more energy to remove, as they are more strongly attracted to the nucleus. The electrons in the innermost shell, being the closest to the nucleus, are the hardest to remove. From the large increases in the ionisation energy as we move from the outermost shell inwards, we can deduce the total number of shells in the atom. Also, the number of data points in each section of the plot that corresponds to one shell is equal to the number of electrons in that shell.

From the graph, we deduce that the element X has three shells and three electrons in the outer shell, which means X is aluminium.

(iii) $Al^{2+}_{(g)} \rightarrow Al^{3+}_{(g)} + e^-$

(iv) The log of the ionisation energy is used in order to fit the large range of values on the scale.

Background

The amount of energy required to remove an electron from an atom depends on how strongly the electron is attracted to the nucleus. The strength of this attraction depends on two factors

1. The atomic size (the distance of the electron from the nucleus)
2. The amount of positive charge in the nucleus (the nuclear charge)

As the electrons in a given shell are removed one by one, the ion becomes increasingly more positive, making it harder to remove more electrons due to the increased attraction with the nucleus. This increased attraction means that the nucleus is able to exert more pull on the electrons, thus bringing these closer to the nucleus (making the ion smaller), which makes it even harder to remove more electrons. So, ionisation energy increases from one electron to the next in the same shell due to the increased charged and smaller size of the positive ion.

As we work from one shell inwards, the number of shells remaining in the ion decreases, and the attraction between the remaining electrons and the nucleus increases rapidly. Each of the large increases seen in the ionisation energy graph corresponds to the electron being removed from a new shell that is significantly closer to the nucleus; hence we can deduce the number of shells in the atom (from the large increases in ionisation energy) as well as the number of electrons in each shell (from the number of data points in each section of the graph), as shown in Figure 1.3.

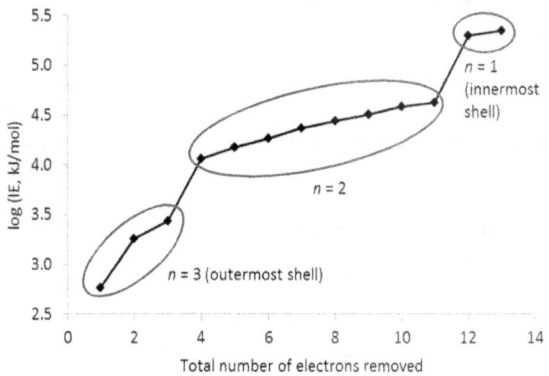

Figure 1.3: Using ionisation energy to deduce the electron arrangement of element X

The actual values of ionisation energy (IE) are plotted in Figure 1.4 using the normal linear scale. Compare this to the logarithmic plot in Figure 1.2. Note that the general shape of the logarithmic plot is not identical to the shape of the plot of the actual IE values, since the logarithm function is not a linear function. In fact, it is this feature of the logarithm function that makes it a useful tool for reducing the scale of the plot of a large range of values. It is easier to recognise the three regions of the plot using the logarithmic scale in Figure 1.2 than using the normal linear scale in Figure 1.4.

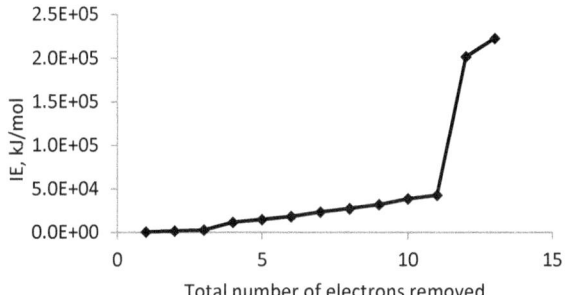

Figure 1.4: Successive ionisation energies of element X on a linear scale

---

Question 1.19

(i) Write the full electron structures of the potassium ion K⁺ and the argon atom Ar.

(ii) The first ionisation energy of Argon is 1521 kJ/mol, while the second ionisation energy of potassium is 3051 kJ/mol. Explain why the second ionisation energy of potassium is higher than the first ionisation energy of argon.

Answer

(i) K⁺ $1s^2, 2s^2, 2p^6, 3s^2, 3p^6$

   Ar $1s^2, 2s^2, 2p^6, 3s^2, 3p^6$

(ii) K⁺ and Ar have the same number of shells. However, the ion K⁺ is smaller than the argon atom and has a higher nuclear charge. Therefore, the attraction between the outer electrons and the nucleus in K⁺ is stronger than that in Ar, and so more energy is needed to remove the electrons.

Background

The potassium atom K (atomic number 19) has the electronic structure $1s^2, 2s^2, 2p^6, 3s^2, 3p^6, 4s^1$ with four electron shells and one electron in the outermost shell. When the potassium atom loses its outer electron, it forms the ion K⁺, which has three electron shells. The atom of argon Ar (atomic number 18) also has three electron shells. However, since the nucleus of K⁺ is more positive than that of Ar (since K⁺ has more protons), it exerts a stronger attraction on the electrons, thus pulling them closer towards it. This makes K⁺ smaller in size than Ar. Therefore, it requires more energy to remove an electron from K⁺ than from Ar.

Question 1.20

Arrange the following atoms in order of increasing first ionisation energy: O, F, S. Explain your answer.

Answer

(lowest) S < O < F (highest)

The first ionisation energy of fluorine is higher than that of oxygen because, while fluorine and oxygen have the same number of shells, fluorine has a higher nuclear charge and a smaller atomic size.

The first ionisation energy of sulphur is lower than that of oxygen because sulphur has one more inner shell, and therefore the outer electrons in sulphur are more shielded from the nucleus and so they are easier to remove.

Background

Ionisation energy increases as the nuclear charge increases and the atomic size decreases. Fluorine and oxygen have the same number of shells or energy levels (they are in the same period in the periodic table). However, as the fluorine atom (atomic number 9) has one more proton in its nucleus compared to the oxygen atom (atomic number 8), its nuclear charge is higher, and so its size is smaller. This results in a stronger attraction between the outer electrons and the nucleus in fluorine; hence the higher value of the first ionisation energy.

The atom of sulphur has three shells while that of oxygen has two. The outermost electron in the sulphur atom is more shielded from the nucleus compared with the outermost electron in the oxygen atom. Therefore, less energy is needed to remove the outer electron from a sulphur atom than from an oxygen atom.

So, to summarize, ionisation energy increases as

1. The number of shells decreases
2. The nuclear charge increases for the same number of shells

2. AMOUNT OF SUBSTANCE

Question 2.1

Define the term relative atomic mass.

Answer

$$Relative\ atomic\ mass = \frac{average\ mass\ of\ one\ atom\ of\ an\ element}{\frac{1}{12}\ the\ mass\ of\ one\ atom\ of\ carbon - 12}$$

Background

Carbon-12 (^{12}C) is used as the baseline for comparing the relative masses of atoms. One unit on the scale of relative atomic mass is equal to one twelfth of the mass of one ^{12}C atom. On this scale, the relative mass of an atom of ^{12}C is exactly 12 (i.e. 12.0000…, etc.). For example, one ^{24}Mg atom weighs twice as much as a ^{12}C atom, and since ^{12}C weighs 12 units, ^{24}Mg weighs 12×2 = 24 units, so the relative atomic mass (A_r) of ^{24}Mg is 24.

Since most elements have different isotopes, the relative atomic mass of an element is calculated taking into account all its naturally occurring isotopes. For example, chlorine has two isotopes, ^{35}Cl and ^{37}Cl. The *average* mass of these two isotopes relative to one twelfth of the mass of a ^{12}C atom is 35.5. This average is calculated using the natural abundance of the two isotopes (see Question 1.10).

Question 2.2

(i) Define the term *relative molecular mass*.

(ii) Calculate the relative molecular mass of Na₂CO₃.

Answer

(i)
$$Relative\ molecular\ mass = \frac{average\ mass\ of\ one\ molecule}{\frac{1}{12}\ the\ mass\ of\ one\ atom\ of\ carbon - 12}$$

(ii) $M_r\ (Na_2CO_3) = (23.0 \times 2) + 12.0 + (16.0 \times 3) = 106.0$

Background

The mass of a molecule is calculated relative to one twelfth of the mass of one atom of ^{12}C. To calculate the relative molecular mass of a formula, simply add up the relative atomic masses of all the atoms present in the molecule. The formula of the molecule states the type and number of atoms that make up the molecule. So, for example, the molecule of chlorine Cl_2 is made up of two chlorine atoms, so the relative molecular mass of Cl_2 is $M_r = 35.5 \times 2 = 71.0$. Note that if the relative atomic masses are given to one decimal place, the final answer should be calculated to one decimal place.

Question 2.3

Define the *Avogadro constant* and explain how it is related to the mole.

Answer

The Avogadro constant is the number of atoms in exactly 12 grams of ^{12}C.

The mole is the amount of substance containing a number of particles equal to the Avogadro constant.

Background

Since the mass of a single atom is exceedingly small, chemists use the mass of large numbers of atoms when performing calculations. It was found that the number of carbon atoms in exactly 12 g of ^{12}C is 6.022×10^{23}. This number is the Avogadro constant, and it is used as the unit of quantity in chemistry. The amount of substance containing 6.022×10^{23} particles is called the *mole*. So, there is one mole of atoms in 12 g of carbon, two moles in 24 g, three moles in 36 g, and so on.

Hydrogen is 12 times lighter than carbon, so 1 mole of hydrogen atoms has a mass of 1.0 g. Helium is four times heavier than hydrogen (A_r (H) = 1.0, A_r (He) = 4.0). So, if 1 mole of hydrogen atoms weighs 1.0 g, then 1 mole of helium atoms weighs 4.0 grams. We say that the molar mass (i.e. the mass of one mole) of helium atoms is 4.0 g/mol. We conclude that the mass of one mole of atoms of an element is numerically equal to the relative atomic mass of the element in grams. Similarly, the mass of one mole of a molecule is equal to its relative molecular mass. So,

Mass of 1 mole of H atoms = 1.0 g

Mass of 1 mole of O atoms = 16.0 g

Mass of 1 mole of H_2O molecules = $1.0 \times 2 + 16.0 = 18.0$ g

---

Question 2.4

Calculate the number of moles of sodium chloride NaCl in 10.0 grams of this salt.

Answer

$$Number\ of\ moles = \frac{mass}{molar\ mass}, \quad or \quad n = \frac{m}{M_r}$$

$M_r(NaCl) = 23.0 + 35.5 = 58.5$

$n = \dfrac{10.0}{58.5} = 0.171\,moles$

Background

We have seen in Question 2.3 that the mass of 1 mole of helium (the molar mass) is 4.0 g. It follows that the mass of 2 moles of helium is 2×4.0=8.0 g, the mass of 3 moles is 3×4.0=12.0 g, and so on. We can therefore say that

    Mass = number of moles × molar mass

But we have also seen that the molar mass is numerically equal to the relative molecular mass M_r (or to the relative atomic mass in the case of atoms), so

    Mass = number of moles × relative molecular mass

$$m = n \times M_r$$

This formula is very useful. It is used to convert the number of moles into mass and vice versa. We can rearrange it to calculate the number of moles *n* in a given mass *m*:

$$n = \frac{m}{M_r}$$

Question 2.5

Calculate the mass of 0.2 mol of H₂SO₄.

Answer

$M_r(H_2SO_4) = (1.0 \times 2) + 32.1 + (16.0 \times 4) = 98.1$

$m = n \times M_r$
   $= 0.2 \times 98.1 = 19.6\,g$

Question 2.6

(i) State the ideal gas equation.

(ii) Calculate the number of moles present in a volume of 500 cm³ of carbon dioxide gas if the temperature of the gas is 25°C and its pressure is 100 kPa (R = 8.31 J K⁻¹ mol⁻¹).

Answer

(i) $PV = nRT$

(ii) $n = \dfrac{PV}{RT} = \dfrac{100 \times 10^3 \times 500 \times 10^{-6}}{8.31 \times (25 + 273)} = 0.0202\,\text{moles}$

Background

A gas which is assumed to obey the following equation is called an ideal gas:

$$PV = nRT$$

P is the gas pressure in Pa

V is the volume occupied by the gas in m³

n is the amount of gas in moles

R is the ideal gas constant (8.31 J K⁻¹ mol⁻¹)

T is the temperature in K

When using this equation, *remember to use the correct units* (International SI units). All the units used must be consistent with the units of the constant R. The conversion factors are given here:

Pressure: 1 kPa = 10³ Pa

Volume: 1 cm³ = 10⁻⁶ m³

                   1 dm³ = 10⁻³ m³

Temperature: K = °C + 273

In this question, the pressure value 100 kPa is converted to Pa by multiplying by 10³, the volume 500 cm³ is converted to m³ by multiplying by 10⁻⁶, and the temperature 25°C is converted into K by adding 273.

The ideal gas equation can be rearranged to calculate P, V, n or T. Given the gas conditions of pressure, volume and temperature, we can find the number of moles of the gas from

$$n = \frac{PV}{RT}$$

Question 2.7

Calculate the volume in dm^3 occupied by 1 mole of any gas at room temperature (20°C) and pressure (100 kPa), given that R = 8.31 J K^{-1} mol^{-1}.

Answer

$$PV = nRT$$
$$V = \frac{nRT}{P} = \frac{1 \times 8.31 \times (20+273)}{100 \times 10^3}$$
$$= 0.0243\,m^3 = 0.0243 \times 10^3\,dm^3 = 24.3\,dm^3$$

Question 2.8

Calculate the mass of carbon dioxide gas present in a volume of 2.0 dm^3 at a temperature of 15°C and a pressure of 110 kPa (R = 8.31 J K^{-1} mol^{-1}).

Answer

Mass is calculated from the formula

$$m = n \times M_r$$

$$M_r(CO_2) = 12.0 + (16.0 \times 2) = 44.0$$

To calculate the number of moles n, we use the ideal gas equation

$$PV = nRT$$
$$n = \frac{PV}{RT} = \frac{110 \times 10^3 \times 2.0 \times 10^{-3}}{8.31 \times (15+273)} = 0.0919\,moles$$

$$m = 0.0919 \times 44.0 = 4.04\,g$$

Question 2.9

A cylinder of volume 40 dm³ contains 4.6 kg of a gas under a pressure of 1.0×10⁴ kPa and stored at room temperature (20°C). Use this information and the ideal gas equation to confirm that the gas in the cylinder is nitrogen N_2.

Answer

We can confirm the identity of the gas by calculating its relative molecular mass (M_r) using the data provided.

$$PV = nRT$$

$$n = \frac{PV}{RT} = \frac{1.0\times10^4 \times 10^3 \times 40\times10^{-3}}{8.31\times(20+273)} = 164.28 \, moles$$

$$m = n \times M_r$$

$$M_r = \frac{m}{n} = \frac{4.6\times10^3}{164.28} = 28.0$$

28.0 is equal to the relative molecular mass of nitrogen N_2.

Background

The ideal gas equation can be used to calculate the relative molecular mass of gases. In this question, we apply the ideal gas equation to find the number of moles of gas stored under the specified conditions of volume, pressure and temperature; then we apply the formula $m = n \times M_r$ to find M_r. The M_r value obtained is equal to the M_r of nitrogen N_2 (14.0×2 = 28.0), which confirms that the gas is indeed nitrogen.

The units of volume, pressure and temperature have to be converted to SI units. The volume is converted from dm³ to m³ by multiplying by 10⁻³. The pressure is converted from kPa to Pa by multiplying by 10³. The temperature is converted from °C to K by adding 273. Note also that the mass has been converted from kg to g by multiplying by 10³.

Question 2.10

(i) Use the periodic table to calculate the relative molecular mass of

- $C_6H_4NO_2$
- C_9H_{14}

(ii) Calculate the relative molecular masses of the compounds in part (i) based on the following relative atomic masses given to 4 decimal places:

C = 12.0000, H = 1.0078, N = 14.0031, O = 15.9949

(iii) Explain how high resolution mass spectrometry can be used to identify the two compounds in part (i).

Answer

(i) $M_r(C_6H_4NO_2) = (12.0\times6) + (1.0\times4) + 14.0 + (16.0\times2) = 122.0$

      $M_r(C_9H_{14}) = (12.0\times9) + (1.0\times14) = 122.0$

(ii) $M_r(C_6H_4NO_2) = (12.0000\times6) + (1.0078\times4) + 14.0031 + (15.9949\times2) = 122.0241$

      $M_r(C_9H_{14}) = (12.0000\times9) + (1.0078\times14) = 122.1092$

(iii) The high resolution mass spectrometer can measure the relative molecular mass (M_r) of a compound correct to several decimal places. The M_r values of the compounds $C_6H_4NO_2$ and C_9H_{14} measured to four decimal places are different; therefore, the two compounds can be identified by their exact M_r values using a high resolution mass spectrometer.

Background

Since high resolution mass spectrometry can measure relative atomic (and molecular) masses to several decimal places, it can be used to identify compounds by their exact molecular masses. Low resolution mass spectrometry cannot always identify a compound by its M_r. For example, a low resolution mass spectrometer would produce the same M_r value (122) for both $C_6H_4NO_2$ and C_9H_{14}. However, a high resolution spectrometer can measure the exact value of M_r to several decimal places and can therefore be used to distinguish between $C_6H_4NO_2$ and C_9H_{14}. Measured to four decimal places, the M_r of $C_6H_4NO_2$ is 122.0241 while that of C_9H_{14} is 122.1092.

Question 2.11

(i) Define the term *empirical formula*.

(ii) What is the empirical formula of C_6H_6.

(iii) 4.29 g of a solid copper oxide was found to contain 3.81 g of copper and 0.48 g of oxygen. Find the empirical formula of the oxide.

Answer

(i) The empirical formula of a compound is the formula that shows the simplest ratio of the atoms of each element in the compound.

(ii) The empirical formula of C_6H_6 is CH.

(iii)

	Cu	O
mass/g	3.81	0.48
A_r	63.5	16.0
moles	$\dfrac{3.81}{63.5} = 0.06$	$\dfrac{0.48}{16.0} = 0.03$
whole number ratio	$\dfrac{0.06}{0.03} = 2$	$\dfrac{0.03}{0.03} = 1$

Therefore, the empirical formula is Cu_2O.

Background

The molecule C_6H_6 contains 6 carbon atoms and 6 hydrogen atoms. The molecular formula C_6H_6 shows the actual number of C and H atoms in the molecule. However, the ratio 6:6 can be simplified to 1:1, giving the empirical formula CH.

The empirical formula of a compound can be calculated from the mass of each element present in the compound. This is done by finding the number of moles of each of these element. In this question, 3.81 g of copper is converted to moles by dividing by the A_r of copper, and 0.48 g of oxygen is converted to moles by dividing by the A_r of oxygen. The number of moles obtained for each element is then converted to a whole number by dividing by the smallest number of moles obtained (0.03). This gives a whole number ratio of Cu:O = 2:1, so the empirical formula contains 2 atoms of Cu to each atom of O, i.e. Cu_2O.

Question 2.12

(a) State the relationship between the empirical formula and the molecular formula for a given compound.

(b) Alkenes are organic compounds with the general formula C_nH_{2n}.

(i) Show that the empirical formula of any alkene is CH_2.

(ii) Deduce the molecular formula of the alkene with $M_r = 56.0$.

Answer

(a) The molecular formula is the empirical formula multiplied by an integer number.

(b) (i) If we divide the number of carbon atoms and hydrogen atoms in C_nH_{2n} by n, we obtain the empirical formula CH_2.

(ii) $M_r(CH_2) = 14.0$

$$integer = \frac{56.0}{14.0} = 4$$

The molecular formula is therefore C_4H_8.

Background

The empirical formula states the simplest ratio of the atoms present in a molecule. The actual ratio of the number of atoms in the molecule is equal to the simplest ratio multiplied by an integer number. To find this integer, we divide the M_r of the actual formula by the M_r of the empirical formula. For example, a compound has the empirical formula C_2H_4O. The actual M_r of the compound is 88.0 while the M_r of the empirical formula C_2H_4O is 12.0×2+1.0×4+16.0 = 44.0. Now, 88.0/44.0 = 2. So, we multiply the number of atoms of each element in the empirical formula C_2H_4O by 2 to get the molecular formula. This gives $C_4H_8O_2$.

Question 2.13

The complete combustion of 0.56 g of an organic compound containing only carbon, hydrogen and oxygen gave 1.23 g of carbon dioxide and 0.67 g of water.

      (i)   Find the empirical formula of the organic compound.
      (ii)   Find its molecular formula given that its relative molecular mass is 60.

Answer

$M_r(CO_2) = 44.0$

Number of moles of CO_2:

$$n = \frac{m}{M_r} = \frac{1.23}{44.0} = 0.0280$$

$M_r(H_2O) = 18.0$

Number of moles of H_2O:

$$n = \frac{m}{M_r} = \frac{0.67}{18.0} = 0.0372$$

Moles of C in the sample = moles of CO_2 produced by combustion

Moles of C = 0.0280

A_r (C) = 12.0

Mass of C = 0.0280 × 12.0 = 0.336 g

Moles of H in the sample = 2 × moles of H_2O produced by combustion

Moles of H = 0.0372 × 2 = 0.0744

A_r (H) = 1.0

Mass of H = 0.0744 × 1.0 = 0.0744

Mass of oxygen in the sample = 0.56 – (0.336 + 0.0744) = 0.150 g

A_r (O) = 16.0

Moles of O = 0.15/16.0 = 0.0094

	C	H	O
moles	0.0280	0.0744	0.0094
whole number ratio	$\dfrac{0.0280}{0.0094} = 3$	$\dfrac{0.0744}{0.0094} = 7.9 \approx 8$	$\dfrac{0.0094}{0.0094} = 1$

So, the empirical formula is C_3H_8O.

(ii) M_r (C_3H_8O) = 60

The M_r of the compound is also 60. Therefore, the molecular formula is C_3H_8O.

Background

Combustion analysis is a method used to find the empirical formula of an unknown compound. The compound usually contains carbon, hydrogen, sulphur, and oxygen. The method involves burning the compound in excess oxygen (to ensure complete combustion) and then measuring the amounts of the oxides produced. On combustion, each element in the compound will produce an oxide. For example, the number of moles of CO_2 produced will be equal to the number of moles of carbon in the compound because all the carbon in the compound is converted to CO_2 and each mole of CO_2 contains one mole of C. The number of moles of H_2O produced will be equal to half the number of moles of H in the compound because all the hydrogen in the compound is converted to H_2O and each mole of H_2O contains two moles of H. Consider the combustion of propanol:

$$C_3H_8O + \frac{9}{2}O_2 \rightarrow 3CO_2 + 4H_2O$$

According to this balanced equation, the number of moles of CO_2 is equal to the number of carbon atoms in propanol, and the number of moles of water is equal to half the number of hydrogen atoms in propanol.

The moles of carbon and hydrogen are converted into mass using $m = n \times M_r$. Since we know the total mass of the sample and the masses of carbon and hydrogen (from the calculated moles of C and H), we can find the mass of oxygen by difference:

Mass of oxygen = total mass – (mass of carbon + mass of hydrogen)

The mass of oxygen is converted into moles. The empirical and molecular formulae are then calculated as described in Questions 2.11 and 2.12.

Question 2.14

An organic compound containing carbon, hydrogen and oxygen was found to have 40.00% carbon and 6.67% hydrogen. Find its molecular formula if $M_r = 60$.

Answer

	C	H	O
mass/g	40.00	6.67	$100 - (40.00 + 6.67)$ $= 53.33$
A_r	12.0	1.0	16.0
moles	$\frac{40.00}{12.0} = 3.33$	$\frac{6.67}{1.0} = 6.67$	$\frac{53.33}{16.0} = 3.33$
whole number ratio	$\frac{3.33}{3.33} = 1$	$\frac{6.67}{3.33} = 2$	$\frac{3.33}{3.33} = 1$

So, the empirical formula is CH_2O. M_r (CH_2O) = 30. Since the M_r of the molecular formula is 60, the molecular formula is $C_2H_4O_2$.

Background

In this question, the amount of each element in the compound is given as a percentage. Therefore, 100 g of the compound will contain 40.00 g of C, 6.67 g of H and 53.33 g of O (Note: always make sure that all the percentages add up to 100%, including any oxygen that may be present.)

Then we calculate the empirical formula as before, and we get CH_2O. This formula has a relative molecular mass of 30. But the M_r of the molecular formula is 60, which means it contains two units of CH_2O. Therefore, the molecular formula is $C_2H_4O_2$.

Question 2.15

(i) Calculate the number of moles of HCl in 200 cm³ of a 0.2 mol dm⁻³ solution of HCl.

(ii) 200 cm³ of the solution in part (i) was diluted with 300 cm³ of water. Calculate the concentration of the new solution.

(iii) What volume of the new solution contains 2.0 g of HCl?

Answer

(i) $n = C \times V = 0.2 \times \dfrac{200}{1000} = 0.04 \, moles$

(ii) Number of moles in 200 cm³ = 0.04

New volume = 200 + 300 = 500 cm³

Concentration of new solution (after dilution):

$$C = \frac{n}{V} = \frac{0.04}{500 \big/ 1000} = 0.08 \, mol \, dm^{-3}$$

(iii) M_r (HCl) = 36.5

The number of moles in 2.0 g of HCl is

$$n = \frac{m}{M_r} = \frac{2.0}{36.5} = 0.0548 \, moles$$

The volume of solution that contains 0.0548 moles is

$$V = \frac{n}{C} = \frac{0.0548}{0.08} = 0.685 \, dm^{-3} = 685 \, cm^{-3}$$

Background

(i) A solution is made by dissolving a solute substance in a solvent, usually water. The concentration of the solution is defined as the amount of solute dissolved in a given volume of solution. For example, a solution of HCl with concentration 0.5 mol dm^{-3} contains 0.5 moles in every 1 dm^3 of solution. The concentration C in mol dm^{-3}, also called the molarity, is given by

$$C = \frac{n}{V}$$

where n is the number of moles of the solute and V is the volume of the solvent in dm^{-3} (litres). It is common to express the volume in cm^3. To convert the volume from cm^3 to dm^3, we divide by 1000.

The number of moles is calculated by rearranging the equation above:

$$n = C \times V$$

(ii) When a solution is diluted, the number of moles of the solute remains the same but the volume of the solution is increased (usually by adding water). So, to calculate the new concentration after dilution, divide the number of moles in the original solution by the new volume (volume of original solution + volume of water added) in dm^3.

(iii) Volume is calculated by rearranging the equation above:

$$V = \frac{n}{C}$$

We know the concentration of the solution (C = 0.08 mol dm^{-3}), but we don't know the number of moles n. We can find the number of moles in 2.0 g of HCl using the formula

$$n = \frac{m}{M_r}$$

Question 2.16

A solution of NaOH is made up by dissolving 10.0 g of NaOH in 1.0 dm^3 of water. Find the concentration of this solution.

Answer

$$n = \frac{m}{M_r} = \frac{10.0}{40.0} = 0.25 \, \text{moles}$$

$$C = \frac{n}{V} = \frac{0.25}{1.0} = 0.25 \, \text{mol dm}^{-3}$$

Question 2.17

You are required to prepare 500 cm^3 of a KOH solution of concentration 0.1 mol dm^{-3}. What mass of solid KOH in grams do you need in order to prepare this solution?

Answer

$$n = C \times V = 0.1 \times \frac{500}{1000} = 0.05 \, \text{moles}$$

$M_r \, (\text{KOH}) = 39.1 + 16.0 + 1.0 = 56.1$

$m = n \times M_r = 0.05 \times 56.1 = 2.81 \, \text{g}$

Background

To find the mass required to prepare the solution, we use $m = n \times M_r$. But since the number of moles n is unknown, we need to calculate it using $n = C \times V$. Remember to convert the volume from cm^3 to dm^3 by dividing by 1000.

Question 2.18

(i) Calculate the mass of water produced when 100 g of ethane C_2H_6 undergoes complete combustion in excess oxygen.

(ii) Calculate the volume in dm^3 of carbon dioxide gas produced in this reaction at a temperature of 35°C and a pressure of 100 kPa.

Answer

$$C_2H_6 + \tfrac{7}{2}O_2 \rightarrow 2CO_2 + 3H_2O$$

$M_r \, (C_2H_6) = 30$

$$n = \frac{m}{M_r} = \frac{100}{30} = 3.333 \, \text{moles of } C_2H_6$$

Molar ratio C_2H_6:H_2O = 1:3

$n\,(H_2O) = 3 \times 3.333 = 10$
$M_r \, (H_2O) = 18$
$m = n \times Mr = 10 \times 18 = 180 \, \text{g}$

(ii) Molar ratio $C_2H_6:CO_2 = 1:2$

$n(CO_2) = 2 \times 3.333 = 6.667$

$PV = nRT$

$V = \dfrac{nRT}{P} = \dfrac{6.667 \times 8.31 \times (35 + 273)}{100 \times 10^3}$

$= 0.171 \, m^3 = 171 \, dm^3$

Background

(i) Questions dealing with reacting masses are usually solved in the same way as follows. You will be given the mass of one of the reactants or products, and you will be asked to calculate the mass of another reactant or product. Let's assume that we know the mass of species A (one of the reactants or products), and we want to find the mass of species B. First, we convert the mass of A into moles of A using $n = \dfrac{m}{M_r}$. The number of moles of B is then calculated by multiplying the number of moles of A by the molar ratio of B:A, which we obtain from the balanced equation of the reaction. We now convert the moles of B into mass using $m = n \times M_r$. This procedure can be summarised as follows:

$$\text{Mass of A} \rightarrow \text{Moles of A} \rightarrow \text{Moles of B} \rightarrow \text{Mass of B}$$

Remember to state any relative molecular mass M_r that you use in the calculations. For example, in the above question, $M_r \, (C_2H_6) = 30$.

(ii) The volume of a gas is calculated using the ideal gas equation $PV = nRT$. The pressure and temperature are known, but we have to calculate the number of moles of CO_2. We use the number of moles of C_2H_6 calculated in part (i) to find the number of moles of CO_2 using the molar ratio $C_2H_6:CO_2 = 1:2$. This means that the number of moles of CO_2 is double that of C_2H_6.

Question 2.19

Sulphur trioxide SO_3 is used in the production of sulphuric acid. SO_3 is made by the oxidation of sulphur dioxide SO_2 according to the equation

$$SO_2 + \tfrac{1}{2}O_2 \rightarrow SO_3$$

(a) **Calculate the mass of SO_2 required for the production of 1 kg of SO_3.**
(b) **When the reaction was carried out at a pressure of 2 atmospheres (200 kPa), it was found that 1 kg of the product gas SO_3 occupied a volume of 400 dm^3. Calculate the temperature of the reaction in degrees Celsius. (The gas constant $R = 8.31$ J mol^{-1} K^{-1}).**

Answer

(a) M_r (SO_3) = 80.1

$$n = \frac{m}{M_r} = \frac{1000}{80.1} = 12.48 \text{ moles of } SO_3$$

Molar ratio $SO_2:SO_3$ = 1:1

$n(SO_2) = 12.48$
$M_r(SO_2) = 64.1$
$m = n \times M_r = 12.48 \times 64.1 = 800.0 \text{ g}$

(b) From the ideal gas equation

$$T = \frac{PV}{nR} = \frac{200 \times 10^3 \times \left(400 / 1000\right)}{12.48 \times 8.31}$$
$$= 771 \text{K} = 771 - 273 = 498°\text{C}$$

Question 2.20

Thermal decomposition of calcium carbonate $CaCO_3$ produces calcium oxide CaO and carbon dioxide CO_2. Write a balanced equation for the reaction and calculate the mass of calcium oxide produced by the decomposition of 1 kg of calcium carbonate.

Answer

$$CaCO_3 \rightarrow CaO + CO_2$$

$M_r(CaCO_3) = 100.1$
$$n = \frac{m}{M_r} = \frac{1000}{100.1} = 9.99 \text{ moles}$$

Molar ratio $CaCO_3:CaO$ = 1:1

$n(CaO) = n(CaCO_3) = 9.99$

$M_r(CaO) = 56.0$

$$m = n \times M_r = 9.99 \times 56.0 = 559.4 \text{ g}$$

Question 2.21

A sample of solid potassium nitrate KNO_3 was heated strongly until fully decomposed into potassium nitrite KNO_2 and oxygen gas.

(i)   Write a balanced equation for the reaction taking place, including state symbols.

(ii)   At 300 K and 101 kPa, the gas produced occupied a volume of 0.8 dm³. Use this information and the balanced equation from your answer to part (i) to calculate the mass of the potassium nitrate sample. (The gas constant $R = 8.31$ J mol^{-1} K^{-1}).

Answer

(i) $KNO_{3(s)} \rightarrow KNO_{2(s)} + \frac{1}{2}O_{2(g)}$

(ii) We use the ideal gas equation to find the moles of O_2:

$PV = nRT$

$$n = \frac{PV}{RT} = \frac{101 \times 10^3 \times 0.8 \times 10^{-3}}{8.31 \times 300} = 0.0324 \, \text{moles}$$

Molar ratio $KNO_3:O_2$: = 2/1 = 2

Therefore, moles of KNO_3 = 0.0324 × 2 = 0.0648 moles

M_r (KNO_3) = 101.1

$$m = n \times M_r = 0.0648 \times 101.1 = 6.55 \, \text{g}$$

Background

(i) The thermal decomposition of potassium nitrate is similar to that of calcium carbonate (see Question 2.20). Note that the balanced equation can be written in two ways

$KNO_{3(s)} \rightarrow KNO_{2(s)} + \frac{1}{2}O_{2(g)}$

$2KNO_{3(s)} \rightarrow 2KNO_{2(s)} + O_{2(g)}$

(ii) The temperature, pressure and volume of oxygen produced are given. Therefore, we can use the ideal gas equation to find the number of moles of oxygen. From the balanced equation, the number of moles of KNO_3 equals 2 × the number of moles of O_2. We can then calculate the mass of KNO_3 using the formula $m = n \times M_r$.

Question 2.22

Magnesium reacts with aqueous hydrochloric acid HCl producing magnesium chloride $MgCl_2$ and hydrogen gas.

    (i)   Write a balanced equation for the reaction.

    (ii)   2.43 g of magnesium was added to 200 cm^3 of a hydrochloric acid solution. It was observed that all the magnesium reacted with the acid and that the resulting solution was neutral (no acid was left). Calculate the number of moles of HCl in the solution, and hence find the concentration of the acid solution.

    (iii)   Calculate the mass of magnesium chloride produced.

    (iv)   Calculate the volume of hydrogen gas produced in the reaction at 293 K and 100 kPa. ($R = 8.31$ J mol^{-1} K^{-1})

Answer

(i) $Mg + 2HCl \rightarrow MgCl_2 + H_2$

(ii) M_r (Mg) = 24.3

Number of moles of Mg : $n = \dfrac{mass}{M_r} = \dfrac{2.43}{24.3} = 0.1 \, moles$

From the balanced equation, the molar ratio HCl/Mg = 2/1. Therefore,

Number of moles of HCl = 2 × 0.1 = 0.2 moles

$$C = \frac{n}{V} = \frac{0.2}{200/1000} = 1.0 \, mol \, dm^{-3}$$

(iii) From the balanced equation, the molar ratio $MgCl_2$/Mg: = 1/1. Therefore,

Number of moles of $MgCl_2$ = 0.1 moles

M_r ($MgCl_2$) = 95.3

Mass of $MgCl_2$ $m = n \times Mr = 0.1 \times 95.3 = 9.53$ g

(iv) We use the ideal gas equation to calculate the volume of H_2 gas

$PV = nRT$

$V = \dfrac{nRT}{P}$

From the balanced equation, the molar ratio H_2/Mg = 1/1. Since n (Mg) = 1, then n (H_2) = 1. Then:

$$V = \frac{0.1 \times 8.31 \times 293}{101 \times 10^3} = 2.41 \times 10^{-3} \, m^3 \text{ or } 2.41 \, dm^3$$

Question 2.23

In an experiment, a magnesium ribbon was added to a flask containing 100 cm³ of hydrochloric acid solution of concentration 0.200 mol dm⁻³. The flask was connected to a syringe in order to measure the volume of the hydrogen gas produced by the reaction of the magnesium ribbon with the acid. When the volume of the gas in the syringe reached 180 cm³, it was found that no magnesium was left. The experiment was performed at a temperature of 25°C and a pressure of 100 kPa.

(i)   Write a full balanced equation for the reaction taking place.

(ii)   Write an ionic equation for the reaction.

(iii)   Calculate the number of moles of hydrogen produced under the experimental conditions stated above. ($R = 8.31$ J mol⁻¹ K⁻¹)

(iv)   Use your answer in part (iii) to calculate the mass of the magnesium ribbon used in the experiment.

(v)   Find the number of moles of HCl in 100 cm³ of 0.2 mol dm⁻³ solution, and hence determine whether the hydrochloric acid used in the experiment was present in excess.

Answer

(i) $Mg + 2HCl \rightarrow MgCl_2 + H_2$

(ii) $Mg + 2H^+ \rightarrow Mg^{2+} + H_2$

(iii) Since the volume of the gas and the experimental conditions of temperature and pressure are known, we can use the ideal gas equation to find the number of moles

$PV = nRT$

$$n = \frac{PV}{RT} = \frac{100 \times 10^3 \times 180 \times 10^{-6}}{8.31 \times (25 + 273)} = 7.27 \times 10^{-3} \text{ moles}$$

(iv) From the balanced equation, the molar ratio Mg/H₂ = 1/1. Therefore,

Number of moles of Mg = 7.27 × 10⁻³ moles

A_r (Mg) = 24.3

Mass of Mg : $m = n \times A_r = 7.27 \times 10^{-3} \times 24.3 = 0.177$ g

35

(v) $n = C \times V = 0.2 \times \frac{100}{1000} = 0.02$ moles

To find the number of moles of HCl reacted with Mg:

From the balanced equation, the molar ratio HCl/Mg = 2/1. Therefore, the number of moles of HCl reacted is $2 \times (7.27 \times 10^{-3}) = 0.0145$ moles. This is smaller than the total number of moles of HCl that was present in the solution (0.02 moles). Therefore, the hydrochloric acid is present in excess.

Background

(ii) Ionic compounds are formed between metals and non-metals. For example, magnesium chloride $MgCl_2$ is ionic because magnesium is a metal and chlorine is a non-metal. In aqueous solution, an ionic compound that is soluble in water will be present in the form of ions. So $MgCl_2$ dissociates in water to give one ion of Mg^{2+} and two ions of Cl^-. The metal ion is almost always positively charged and the non-metal ion is negatively charged. The charge on the ion of an element is equal to the valence of the element. The valence is the number of electrons the atom loses, gains, or shares in its compounds. For example, a magnesium atom has 2 electrons in its outer shell, and so it tends to lose 2 electrons, thus forming the ion Mg^{2+}. A chlorine atom has 7 electrons in its outer shell, and so it tends to gain an electron, thus forming the ion Cl^-. The valences of the elements from Group 1 to Group 7 of the periodic table are given here:

Group	1	2	3	4	5	6	7
Valence	1	2	3	4	3	2	1

For example, potassium, a metal in Group 1, has a valence of 1, so it forms the positive ion K^+. Oxygen is a non-metal in Group 6, so it has a valence of 2 and it forms the negative ion O^{2-}.

In addition to ionic compounds, strong acids also dissociate in water to produce two types of ions, one of which is always H^+. For example, hydrochloric acid dissociates in aqueous solution according to the equation

$$HCl \rightarrow H^+ + Cl^-$$

Similarly, sulphuric acid dissociates as follows:

$$H_2SO_4 \rightarrow 2H^+ + SO_4^{2-}$$

Returning to the question above, the reaction involves two compounds that are present in the solution as ions: HCl (because it's a strong acid) and $MgCl_2$ (because it's an ionic compound). So, we write their formulae in the ionic form as follows:

$$Mg + 2(H^+ + Cl^-) \rightarrow (Mg^{2+} + 2Cl^-) + H_2$$

36

Note that there are two Cl⁻ ions on each side of the equation, which means that the Cl⁻ ions do not undergo any change in this reaction. Such ions are called *spectator* ions, and since they don't take part in the reaction, they are cancelled out in the ionic equation:

$$Mg + 2H^+ \rightarrow Mg^{2+} + H_2$$

This ionic equation tells us that the magnesium atom loses two electrons, while each of the two hydrogen ions gains one electron.

(iii) We use the ideal gas equation to find the number of moles of hydrogen produced. Remember to convert the units of volume, pressure, and temperature to m^3, Pa, and K respectively.

(iv) The balanced equation of the reaction tells us that 1 mole of Mg produces 1 mole of H_2; therefore, the number of moles of Mg must be equal to the number of moles of H_2 calculated in part (iii). The mass of Mg is then calculated from the number of moles ($m = n \times A_r$).

(v) We find the total number of moles of HCl in the solution using $n = C \times V$. The number of moles of HCl that reacted with Mg is equal to 2 × the number of moles of Mg (from the balanced equation of the reaction). The calculations show that the number of moles of HCl that reacted with Mg is less than the total number of moles of HCl that were present in the original solution. This means that the original solution contained more than enough acid to react with all the magnesium, and the hydrochloric acid was therefore in excess.

Question 2.24

30.0 cm³ of a solution of sulphuric acid H_2SO_4 of unknown concentration was neutralised by 22.5 cm³ of a 0.100 mol dm⁻³ solution of sodium hydroxide NaOH. The reaction produced sodium sulphate Na_2SO_4 and water.

(i) Write a balanced equation for the neutralisation reaction.
(ii) Calculate the number of moles of NaOH used in the titration, and hence deduce the number of moles of H_2SO_4 in 30.0 cm³ of the acid solution.
(iii) Use your answer in part (ii) to calculate the concentration of the sulphuric acid solution.

Answer

(i) $\qquad H_2SO_4 + 2NaOH \rightarrow Na_2SO_4 + 2H_2O$

(ii) $\qquad n = C \times V = 0.100 \times {22.5}/{1000} = 2.25 \times 10^{-3} \text{ moles}$

We know from the balanced equation in (i) that the number of moles of H_2SO_4 is half that of NaOH, so the number of moles of H_2SO_4 is

$$n = \frac{2.25 \times 10^{-3}}{2} = 1.13 \times 10^{-3} mol$$

(iii)

$$C = \frac{n}{V} = \frac{1.13 \times 10^{-3}}{30/1000} = 0.038 \; mol \; dm^{-3}$$

Background

Neutralisation is the reaction of an acid with an alkali. Titration is a technique used to determine the concentration of an acid or an alkali solution using neutralisation. Suppose we want to find the concentration of A by titrating it with B. We take a given volume of A (V_A) and titrate it using a solution of B of a given concentration (C_B). If we know the volume of B required for complete neutralisation (V_B), then we can calculate the number of moles of B used in the reaction (n_B) using the formula $n = C \times V$. Now that we know n_B, we can use the molar ratio of A:B (from the balanced equation of the reaction) to determine the number of moles of A (n_A). Having found n_A, we can now calculate the concentration of A using $C = n_A/V_A$.

Question 2.25

20 cm³ of 0.100 mol dm⁻³ HCl solution are added to 25 cm³ of NaOH solution of concentration 0.100 mol dm⁻³.

    (i)   Write a full balanced equation for the reaction of HCl with NaOH.

    (ii)   Write an ionic equation for the reaction.

    (iii)   Would there be any NaOH left when the reaction finished? Explain your answer.

Answer

(i) $HCl + NaOH \rightarrow NaCl + H_2O$

(ii) $H^+ + OH^- \rightarrow H_2O$

(iii) The number of moles of HCl:

$$n = C \times V = 0.100 \times \frac{20}{1000} = 2.00 \times 10^{-3} \; moles$$

The number of moles of NaOH required to completely react with 2.00×10^{-3} moles of HCl is also 2.00×10^{-3}, since the molar ratio NaOH:HCl is 1/1.

The number of moles of NaOH present in the solution is:

$$n = C \times V = 0.100 \times \frac{25}{1000} = 2.50 \times 10^{-3} \text{ moles}$$

The solution contains more NaOH than is required for complete neutralisation of HCl. Therefore, there will be some NaOH left unreacted.

Background

(ii) HCl is an acid that dissociates to form H^+ and Cl^- ions. NaOH is an ionic compound (and a strong alkali) that dissociates to form Na^+ and OH^- ions. (For more detail, refer to Question 2.23, part (ii)).

(iii) The number of moles of NaOH present in the solution (2.50×10^{-3}) is greater than the number of moles of NaOH reacted (2.00×10^{-3}). Therefore, NaOH is present in excess.

Question 2.26

Calcium carbonate reacts with hydrochloric acid according to the equation

$$CaCO_3 + 2HCl \rightarrow CaCl_2 + CO_2 + H_2O$$

A 1.64 g sample of impure calcium carbonate $CaCO_3$ was added to 125 cm³ of 0.250 mol dm⁻³ hydrochloric acid. The unreacted acid was titrated using sodium hydroxide of concentration 0.200 mol dm⁻³. The titration required 19.2 cm³ of the sodium hydroxide solution for complete neutralisation.

(i)   Calculate the number of moles of HCl in 125 cm³ of 0.250 mol dm⁻³ hydrochloric acid solution.
(ii) Explain why an excess of HCl was used in the reaction with $CaCO_3$.
(iii) Calculate the number of moles of NaOH required for neutralisation.
(iv) Calculate the mass of $CaCO_3$ in the sample, and hence deduce the percentage purity of $CaCO_3$.
(v)   The carbon dioxide gas produced by the reaction occupied a volume of 350 cm³ at a pressure of 100 kPa and a temperature T. State the ideal gas equation and use it to calculate the temperature T. ($R = 8.31$ J mol⁻¹ K⁻¹)

Answer

(i) $n = C \times V = 0.250 \times \frac{125}{1000} = 0.0313 \text{ moles}$

(ii) Excess HCl was used to ensure that all the calcium carbonate in the sample reacted with the acid.

(iii) $n = C \times V = 0.200 \times \frac{19.2}{1000} = 0.00384$ moles

(iv) Molar ratio of HCl:NaOH is 1/1, therefore,

Number of moles of HCl reacted with NaOH = 0.00384 moles

Number of moles of HCl reacted with $CaCO_3$ = 0.0313-0.00384 = 0.0275 moles

Molar ratio $CaCO_3$:HCl = 1/2, therefore,

Number of moles of $CaCO_3$ = $0.0275 \times \frac{1}{2} = 0.0138$ moles

$M_r(CaCO_3) = 100.1$

$m = n \times M_r = 0.0138 \times 100.1 = 1.38 \, g$

$$\text{Percentage purity} = \frac{\text{Mass of } CaCO_3}{\text{Total mass of sample}} \times 100$$

$$= \frac{1.38}{1.64} \times 100 = 84.1\%$$

(v) $\qquad PV = nRT$

Number of moles of CO_2 is equal to the number of moles of $CaCO_3$ (1/1 ratio).

$$T = \frac{PV}{nR} = \frac{100 \times 10^3 \times 350 \times 10^{-6}}{0.0138 \times 8.31} = 305 \, K$$

Background

In addition to $CaCO_3$, the sample of calcium carbonate contains some impurities. To determine the amount of pure $CaCO_3$ in the sample, the sample is reacted with excess hydrochloric acid HCl. If we find the amount of acid reacted, we can then determine the amount of $CaCO_3$. The amount of acid reacted is found by titrating the unreacted acid with sodium hydroxide.

First, we add the calcium carbonate sample to a solution of hydrochloric acid to dissolve all the calcium carbonate. The impurities do not react with HCl, and so only $CaCO_3$ in the sample will react. But since we do not know the exact amount of HCl required for dissolving the calcium carbonate completely, an excess of the acid is used to ensure that all of the carbonate in the sample reacted completely. Since the acid is in excess, the resulting solution will contain some unreacted acid. The amount of unreacted HCl can be determined by titrating the resulting solution with NaOH.

From the concentration and volume of the NaOH solution used in the titration, we can find the number of moles of NaOH reacted ($n = C \times V$). Then, using the molar ratio HCl:NaOH=1/1, we deduce the number of moles of HCl that reacted with NaOH (let's call it X). This amount, X, is the excess amount of acid left after the first reaction with $CaCO_3$.

From the concentration and volume of the original HCl solution, we can find the total number of moles of HCl before any reaction. The amount of HCl reacted with $CaCO_3$ is then calculated as the difference between the total amount of HCl in the original solution and the amount of HCl left (X).

The number of moles of $CaCO_3$ in the sample is equal to half the number of moles of HCl reacted (molar ratio $CaCO_3:HCl = 1/2$). We then convert the number of moles of $CaCO_3$ into mass. The percentage purity of $CaCO_3$ is the percentage of pure $CaCO_3$ in the impure sample:

$$\text{Percentage purity} = \frac{\text{Mass of } CaCO_3}{\text{Total mass of sample}} \times 100$$

Question 2.27

(a) An organic compound contains 15.9% of carbon, 18.5% of nitrogen, and 63.4% of oxygen, the rest being hydrogen. Find the empirical formula of this compound.

(b) Nitroglycerine is an explosive with the molecular formula $C_3H_5N_3O_9$. When detonated, nitroglycerine decomposes according to the following equation

$$4C_3H_5N_3O_{9(l)} \rightarrow 12CO_{2(g)} + 10H_2O_{(g)} + 6N_{2(g)} + O_{2(g)}$$

30.0 g of nitroglycerine were placed in a sealed container of volume 2.00×10^{-3} m^3 and detonated. The temperature in the container after detonation reached 1000 K. Use the ideal gas equation to calculate the pressure in the container after detonation. ($R = 8.31$ J mol^{-1} K^{-1})

Answer

(a)

	C	H	N	O
mass/g	15.9	100- (15.9+18.5+63.4) = 2.2	18.5	63.4
A_r	12.0	1.0	14.0	16.0
moles	$\frac{15.9}{12.0}=1.325$	$\frac{2.2}{1.0}=2.2$	$\frac{18.5}{14.0}=1.321$	$\frac{63.4}{16.0}=3.963$
÷1.321	$\frac{1.325}{1.321}\approx1.00$	$\frac{2.2}{1.321}=1.67$	$\frac{1.321}{1.321}=1.00$	$\frac{3.963}{1.321}=3.00$
×3	3	5	3	9

Therefore, the empirical formula is $C_3H_5N_3O_9$.

(b) M_r ($C_3H_5N_3O_9$) = 227

Number of moles of $C_3H_5N_3O_9$:

$$n = \frac{m}{M_r} = \frac{30.0}{227} = 0.132 \, \text{moles}$$

From the balanced equation, 4 moles of $C_3H_5N_3O_9$ produce a total of 29 moles of gas; therefore, the number of moles of gas produced after detonation is

$$n\,(\text{gas}) = \frac{29}{4} \times 0.132 = 0.958 \, \text{moles}$$

$$PV = nRT$$
$$P = \frac{nRT}{V} = \frac{0.958 \times 8.31 \times 1000}{2.00 \times 10^{-3}} = 3.98 \times 10^6 \, \text{Pa}$$

Background

(a) The organic compound in this question contains C, H, N, and O. The percentages of C, N, and O are known, and the percentage of H is calculated by difference (H = 100 - C - N - O). We then find the number of moles of each element by dividing its percentage by its relative atomic mass A_r. To obtain a whole number ratio, we divide by the smallest number of moles (1.321). This results in the ratio 1:1.67:1:3. Note that one of the numbers obtained (1.67) is not a whole number. Therefore, we multiply the ratio by a whole number (2, 3, etc.) until we obtain a whole number ratio. We find that when we multiply 1.67 by 3 we get a whole number (5.0), therefore we multiply through by 3. Hence, the empirical formula is $C_3H_5N_3O_9$.

(b) When nitroglycerine is detonated in a sealed container, the reaction generates a large number of gaseous molecules. According to the balanced equation, 4 moles of nitroglycerine produce 29 moles of gas.

We can use the ideal gas equation to calculate the pressure generated by the gases in the container:

$$P = \frac{nRT}{V}$$

While T and V are known, the number of moles n is unknown. Note that n here is the number of moles of the gases produced by the decomposition of nitroglycerine, and not the number of moles of nitroglycerine itself. To find n, we first calculate the number of moles of nitroglycerine from its mass using $n = m/M_r$. Then, from the balanced equation of the reaction we find that the molar ratio of the gases to $C_3H_5N_3O_9$ is 29/4. Therefore, the number of moles of the gases is equal to the number of moles of $C_3H_5N_3O_9$ multiplied by 29/4.

When sulphuric acid is added to a solution containing barium chloride ($BaCl_2$), a white precipitate of barium sulphate ($BaSO_4$) is formed. In an experiment, sulphuric acid was added to a solution containing 4.60 g of $BaCl_2$. The precipitate formed was filtered off, dried, and weighed. The mass of the precipitate was found to be 4.10 g.

    (i)   Write a full balanced equation for the reaction, including state symbols.
    (ii)   Write an ionic equation for the reaction.
    (iii) Calculate the maximum mass of $BaSO_4$ that could theoretically be obtained from 3.0 g of $BaCl_2$, and hence find the percentage yield.
    (iv) Suggest two reasons why the percentage yield of this reaction is less than 100%.

Answer

(i) $BaCl_{2\,(aq)} + H_2SO_{4\,(aq)} \rightarrow BaSO_{4\,(s)} + 2HCl_{(aq)}$

(ii) $Ba^{+2}_{(aq)} + SO^{2-}_{4(aq)} \rightarrow BaSO_{4\,(s)}$

(iii) First, we find the moles of $BaCl_2$ in 4.60 g:

$M_r\,(BaCl_2) = 137.3 + 35.5 \times 2 = 208.3$

$n = \dfrac{m}{M_r} = \dfrac{4.60}{208.3} = 0.0221\,moles$

Molar ratio $BaSO_4/BaCl_2$ = 1/1, therefore

$n\,(BaSO_4) = 0.0221$ moles
$Mr\,(BaSO_4) = 137.3 + 32.1 + 16.0 \times 4 = 233.4$
$m = n \times M_r = 0.0221 \times 233.4 = 5.16\,g$

$\text{Percentage yield} = \dfrac{\text{Actual mass}}{\text{Theoretical mass}} \times 100 = \dfrac{4.10}{5.16} \times 100 = 79.5\%$

(iv) The yield is less than 100% because:

    (1) The reaction may be incomplete
    (2) Some of the product may be lost during filtration

Background

The equation of the reaction predicts that 1 mole of $BaCl_2$ will produce 1 mole of $BaSO_4$. However, this is in theory only. In practice, the amount of product obtained will always be less than the theoretical amount predicted by the equation. In other words, the yield will always be less than 100%. This can be due to any of the following reasons:

(1) Some reactions are reversible and do not go to completion.

(2) The reactants may undergo side reactions. These are reactions which do not produce the desired product.

(3) Loss of product during separation. In most cases, the separation of the product from the reaction mixture entails some loss of the product. For example, in the above experiment, the barium sulphate precipitate is separated by filtration. When the precipitate is removed from the filter paper, a small amount will inevitably be left on the paper.

(4) Experimental error, such as inaccurate measurements.

Question 2.29

Ibuprofen is a painkiller and anti-inflammatory drug with a relative molecular mass of 206. In a new method of manufacture adopted in the early 1990s, the production of 1 mole of ibuprofen requires 1 mole of each of the following reagents

Reagent	Formula	M_r
2-methylpropylbenzene	$C_{10}H_{14}$	134
Ethanoic anhydride	$C_4H_6O_3$	102
Hydrogen	H_2	2
Carbon monoxide	CO	28

(i) Calculate the atom economy of this process.

(ii) An older process had an atom economy of 40%. Explain why the new method is considered 'greener', i.e. more environmentally friendly, than the old method.

(iii) Theoretically, 1 mole of ibuprofen can be produced from 1 mole of 2-methylpropylbenzene, $C_{10}H_{14}$. Calculate the maximum mass of ibuprofen that can be made starting from 1.0 kg of 2-methylpropylbenzene.

(iv) If the overall yield of the process was 70%, what mass of ibuprofen would actually be produced from 1.0 kg of 2-methylpropylbenzene?

Answer

(i) Mr of Ibuprofen = 206. Total M_r of reactants = 134 + 102 + 2 + 28 = 266

$$\% \text{ Atom economy} = \frac{\text{mass of desired product}}{\text{total mass of reactants}} \times 100 = \frac{206}{266} \times 100 = 77.4\%$$

(ii) The atom economy of the new method (77.4%) is higher than the atom economy of the old method (40%). This means that the new method produces less waste than the old method, and it is therefore greener.

(iii) Number of moles in 1.0 kg (1000 g) of $C_{10}H_{14}$

$$n = \frac{m}{M_r} = \frac{1000}{134} = 7.46 \, \text{moles}$$

Molar ratio ibuprofen/$C_{10}H_{14}$ = 1/1, therefore, the number of moles of ibuprofen must be equal to the number of moles of $C_{10}H_{14}$:

$$n(\text{ibuprofen}) = 7.46$$
$$m = n \times M_r = 7.46 \times 206 = 1.54 \times 10^3 \, \text{g} = 1.54 \, \text{kg}$$

(iv) The actual mass is calculated from the formula

$$\text{Actual mass} = \text{Theoretical mass} \times \frac{\text{yield}}{100}$$
$$= 1.54 \times \frac{70}{100} = 1.08 \, \text{kg}$$

Background

(i) The process of making ibuprofen involves more than one chemical reaction. We do not need to know the details of the reactions taking place in order to answer this question. Given in the question are the reagents used in the manufacture of ibuprofen along with their relative molecular masses. The atom economy is defined as

$$\% \, \text{Atom economy} = \frac{\text{mass of desired product}}{\text{total mass of reactants}} \times 100$$

We calculate the atom economy based on 1 mole of ibuprofen (the desired product). To make 1 mole of ibuprofen, we need 1 mole of each of $C_{10}H_{14}$, $C_4H_6O_3$, H_2 and CO. Therefore, the mass of the desired product is the mass of 1 mole of ibuprofen (M_r = 206), and the total mass of the reactants is the sum of the masses of 1 mole of each of the reagents (total M_r = 134 + 102 + 2 + 28 = 266).

(ii) The atom economy is a measure of the amount of reactants that is converted to the desired product. In the old process, only 40% of the mass of the reactants was converted to ibuprofen, while the rest (60%) was converted to by-products that were disposed of as waste. Disposing of large quantities of wastes can be an environmental problem. The new method has an atom economy of 77.4%, and so it reduces the amount of wastes from 60% to 22.6%. In fact, the only major by-product of this process is ethanoic acid, which itself is a useful product that can be recovered and used for other purposes. Taking this into consideration, only 1% of the reactants is actually wasted, which makes the new process far better than the old one.

(iii) The same method used in previous questions to calculate reacting masses is used here.

(iv) The actual mass produced is always less than the theoretical mass (calculated in part (iii)) and can be calculated by rearranging the formula of the percentage yield:

$$\% \text{ Yield} = \frac{\text{actual mass}}{\text{theoretical mass}} \times 100$$

Actual mass = theoretical mass $\times$ $\text{yield}/100$

Question 2.30

In the Haber process, ammonia is made from nitrogen and hydrogen according to the reversible reaction

$$N_2 + 3H_2 \rightleftharpoons 2NH_3$$

(i) Define the term *atom economy* and explain why the atom economy of this reaction is 100%.

(ii) At a temperature of 500°C and a pressure of 20×10^3 kPa, the yield of this reaction is approximately 20%. Suggest a reason why the yield is very low.

(iii) Comment on the atom economy and percentage yield of this reaction under the above conditions.

Answer

(i) Atom economy $= \dfrac{\text{mass of desired product}}{\text{total mass of reactants}} \times 100$

The atom economy of this reaction is 100% because all the atoms in the reactants are incorporated into the desired product (there are no by-products).

(ii) The yield of this reaction is low because the reaction is reversible.

(iii) The atom economy is very high (100%) and there are no waste products, which means that all the atoms in the reactants that are actually converted are used to make the desired product. However, the percentage yield is very low, which means that only a small percentage of the reactants is actually converted. Most of the hydrogen and nitrogen are left unreacted.

Background

(i) In this reaction, all the atoms in the reactant molecules which are converted by the reaction are incorporated into the desired product. In other words, no by-products are formed. Therefore, 100% of the atoms in the reactants are used to make ammonia (the desired product).

(ii) This reaction is reversible, which means that not all the nitrogen and hydrogen present at the beginning of the reaction will be converted. A significant amount of the reactants will be left unreacted. Therefore, the reaction will not produce the maximum possible amount of ammonia, and the yield is therefore low.

(iii) This reaction only produces ammonia (the desired product), and there are no waste products. This is why the atom economy is 100%. But a good atom economy is not enough to make the process economically viable or profitable. A reasonable yield is also required. This process has a low yield under the conditions stated above, which means that a large amount of the reactant molecules will be left unreacted. For those molecules that *do* react, however, all the atoms are converted into the desired product (i.e. the atom economy is 100%). The yield may be improved by altering the reaction conditions or by recycling the unreacted nitrogen and hydrogen.

Note that there is a significant difference between the yield and the atom economy of a reaction. For example, consider the following generalised reaction

$$A \rightarrow B + C$$

where A is the reactant, B is the desired product, and C is the by-product. If 1 mole of A is present at the beginning of the reaction, the maximum (theoretical) amount of B that could be made is 1 mole. But in practice we obtain, say, 0.7 mole of B. This is because some of A may be left unreacted or may be used up in a side reaction which does not produce B. Also, some of B produced may be lost during purification (the separation of B from the reaction mixture). Therefore, the yield of the reaction is 70%, which means that, for various reasons, we actually obtain 70% of the maximum amount of B that could theoretically be obtained. It also means that only 70% of A is converted into B. 30% of A remains unreacted (or undergoes a different reaction).

The atom economy, on the other hand, is a measure of the percentage of the mass of the atoms in A that are incorporated into the product B, regardless of any side reactions, any unreacted A, or any loss in B during separation. So, if the atom economy is 80%, this means that, of the total mass of A that *is* converted according to the above reaction, 80% is incorporated into B, while the remaining 20% is incorporated into the by-product C.

Question 2.31

Ibuprofen is a painkiller and anti-inflammatory drug. It was originally manufactured from 2-methylpropylbenzene in a six-step process with an atom economy of 40%. In 1992, a new three-step method was used to manufacture ibuprofen from the same starting material. The new process had an atom economy of 77%, with ethanoic acid as the only by-product.

Explain why the new method is more cost-effective and has a lower environmental impact than the original method. Suggest how the new process can be made even more cost-effective.

Answer

The original method had an atom economy of 40%, which means that only 40% of the starting materials was incorporated into ibuprofen, while the remaining 60% was thrown away as waste by-products that may harm the environment. The new method has a much higher atom economy of 77%, which means that 77% of the starting materials is incorporated into ibuprofen and so there are much less waste products.

The new process can be made even more cost-effective by recovering the ethanoic acid, which is a useful by-product, to minimize the amount of wastes.

Question 2.32

The hydroxide of a metal X has the formula XOH and reacts with hydrochloric acid according to the following equation

$$XOH + HCl \rightarrow XCl + H_2O$$

A solution of XOH was prepared by dissolving 1.27 g of the solid hydroxide in 250 cm³ of water. A 25.0 cm³ portion of the resulting solution required 22.5 cm³ of 0.100 mol dm⁻³ hydrochloric acid for complete neutralisation.

(i) To which group (column) in the periodic table does X belong?

(ii) Calculate the number of moles of HCl required for complete neutralisation, and hence deduce the number of moles of XOH in 25.0 cm³ of its solution.

(iii) Use your answer in part (ii) to find the relative molecular mass of XOH, and hence identify the metal X.

Answer

(i) Group 1.

(ii) Number of moles of HCl:

$$n = C \times V = 0.100 \times \frac{22.5}{1000} = 2.25 \times 10^{-3} \text{ moles}$$

Molar ratio XOH/HCl = 1/1; therefore

Number of moles of XOH = 2.25 × 10⁻³ moles in 25 cm³

(iii) The number of moles of XOH in 250 cm^3 = $2.25 \times 10^{-3} \times \dfrac{250}{25} = 0.0225$ moles

$$M_r(\text{XOH}) = \frac{m}{n} = \frac{1.27}{0.0225} = 56.4$$

$Ar(\text{X}) = 56.4 - (16.0 + 1.0) = 39.4$

Therefore, the metal X is potassium K.

Background

The formula of the hydroxide of X is XOH. Since the hydroxide ion (OH$^-$) carries a 1$^-$ charge and the molecule XOH carries no overall charge, the metal ion must carry a 1$^+$ charge to balance the charge of the OH$^-$ ion. The metal X must therefore be in group 1, since the ions of group 1 elements all have a 1$^+$ charge.

The number of moles of HCl is calculated from the volume and concentration of the HCl solution. The number of moles of XOH in 25 cm^3 is equal to the number of moles of HCl since the molar ratio is 1/1 (because one molecule of XOH is required to react with one molecule of HCl to produce one molecule of H$_2$O). To obtain the number of moles of XOH in the original 250 cm^3 solution, we multiply the answer by 250/25, or 10. Since the mass of XOH dissolved in the original solution is known, we can rearrange the formula $m = n \times M_r$ to find the M_r of XOH. From the value of M_r we subtract the relative atomic masses of oxygen and hydrogen to obtain the relative atomic mass of X. The value obtained is 39.4, and the closest relative atomic mass in group 1 to this value is that of potassium (39.1). Therefore, X must be potassium. Note that the relative atomic mass obtained (39.4) is not 100% accurate because it was calculated using experimental measurements, and we know that experimental measurements normally involve some error.

3. BONDING

Question 3.1

(i) Explain the difference between covalent and ionic bonding.

(ii) Write the electron arrangement of the following species:

Na, Na^+, Cl, Cl^-.

(iii) Use your answer in part (ii) to explain how the ionic bond in sodium chloride is formed.

Answer

(i) A covalent bond is formed when two atoms share a pair of electrons. Ionic bonding is the electrostatic attraction between oppositely charged ions formed when electrons are transferred from one atom to another.

(ii)

Na	2,8,1
Na^+	2,8
Cl	2,8,7
Cl^-	2,8,8

(iii) The single outer electron of the sodium atom moves into the outer energy level of the chlorine atom. As a result, sodium forms the positive ion Na^+ while chlorine forms the negative ion Cl^-. The oppositely charged ions Na^+ and Cl^- are attracted to each other by electrostatic forces.

Background

A covalent bond is formed when two non-metal atoms *share* one or more pairs of electrons from their outer energy levels. The shared electrons attract the nuclei of both atoms, thus forming a bond between the two atoms.

An ionic bond is formed when electrons are *transferred* from a metal atom to a non-metal atom so that each atom attains a stable noble gas electron arrangement. The metal atom loses its outer electrons, thus becoming a positively charged ion (a cation). The non-metal atom gains the electrons into its outer energy level, thus becoming a negatively charged ion (an anion). The positive and negative ions attract each other by electrostatic forces forming an ionic bond. For example, when a sodium atom Na encounters a chlorine atom Cl, an electron jumps from Na to Cl, leaving both atoms with filled outer shells. The Na atom that lost an electron now has one less electron than it has protons; it therefore has a net positive charge of +1 (Na^+). The Cl atom that gained an electron now has one more electron than it has protons; it therefore has a net negative charge of -1 (Cl^-).

The general rule is that covalent bonding is formed between two non-metal atoms, while ionic bonding is formed between a metal atom and a non-metal atom. (There are exceptions to this rule. For example, aluminium chloride and beryllium chloride are both covalent.)

Question 3.2

(i) Write the electron structure of each of the following:

Na, Na$^+$, Ne, Cl, Cl$^-$, and Ar.

(ii) Use your answer in part (i) to explain why a sodium atom loses an electron while a chlorine atom gains an electron when sodium chloride is formed.

Answer

(i)

Na	2,8,1
Na$^+$	2,8
Ne	2,8
Cl	2,8,7
Cl$^-$	2,8,8
Ar	2,8,8

(ii) A sodium atom loses one electron to attain the electron structure of the nearest noble gas, Ne. A chlorine atom gains one electron to attain the electron structure of the nearest noble gas, Ar.

Background

Generally, an atom forms chemical bonds in order to attain an electron arrangement which is similar to that of the nearest noble gas in the periodic table. The electron structure of a noble gas is stable because it has a full outer shell of electrons. The sodium atom has the electron arrangement 2,8,1, and by losing the lone electron from the third shell, its electron structure becomes similar to that of neon Ne (2,8). The chlorine atom has the electron arrangement 2,8,7, and by gaining one electron, its electron arrangement becomes similar to that of argon Ar (2,8,8).

Question 3.3

Name the type of bonding present in sodium chloride NaCl and explain how it gives rise to the following properties:

(a) NaCl is a solid at room temperature
(b) NaCl is soluble in water
(c) NaCl conducts electricity when molten or dissolved in water but not when solid
(d) Solid NaCl is brittle

Answer

The bonding in NaCl is ionic.

(a) Solid NaCl has a giant ionic structure with a large number of ionic bonds between the oppositely charged ions Na^+ and Cl^-. It requires a large amount of energy to break these bonds, and so NaCl is solid at room temperature.

(b) NaCl is soluble in water because its ions (Na^+ and Cl^-) can be attracted to water molecules through strong ion-dipole attractions.

(c) NaCl conducts electricity when molten or dissolved in water because in the liquid state the ions are free to move and carry the electric current, while in the solid state they are not free to move.

(d) In solid NaCl, the ions form a lattice of alternating positive and negative ions. When the lattice is given a sharp blow, the ions move in such a way that produces contact between ions with the same charge, thus resulting in repulsion that leads to shattering of the structure.

Background

(a) NaCl is an ionic compound. Ionic compounds form giant ionic lattices in which each ion is attracted to a number of oppositely charged ions, giving rise to a strong structure. A large amount of energy is needed to break the large number of ionic bonds in the lattice, which is why ionic compounds are solid at normal conditions.

(b) Ionic compounds are usually soluble in water because water molecules are polar and can easily be attracted to ions. In a water molecule, the oxygen is partially negative while the hydrogen is partially positive. When an ionic compound is dissolved in water, the solid lattice dissociates and the resulting positive and negative ions form ion-dipole bonds with the water molecules, as shown in Figure 3.1. Molecules that cannot form relatively strong forces of attraction with water molecules are usually insoluble in water.

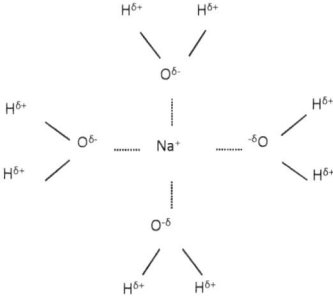

Figure 3.1: ion-dipole attraction forces between a sodium ion and water molecules

(c) If an ionic compound is in the liquid state (molten or dissolved in a solvent), its ions will be able to move freely and conduct electricity (since electricity requires free-moving charged particles). In the solid state, however, the ions are fixed in position within a strong ionic structure, unable to move and conduct electricity.

(d) When an ionic lattice receives a sharp blow, the layers of ions are displaced relative to each other. This displacement brings ions with the same charge closer together (Figure 3.2), resulting in strong repulsive forces that can lead to shattering of the lattice. Ionic lattices are therefore brittle.

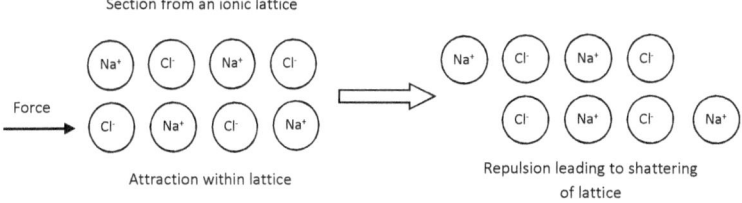

Figure 3.2: Ionic solids are brittle, an applied force will lead to shattering of the lattice.

Note that the important thing in this question is to be able to explain the properties of a substance in terms of its structure and bonding.

Question 3.4

Chlorine has a simple covalent structure. Explain how this structure gives rise to the following properties:

  (a) Chlorine is a gas at room temperature
  (b) Chlorine is a poor conductor of electricity

Answer

(a) Chlorine is a simple covalent molecule with weak intermolecular forces of attraction between its molecules, therefore only a small amount of energy is required to move the molecules apart from each other.

(b) The molecules of chlorine are neutral; therefore, there are no charged particles to carry the electric current.

Background

The bond within a chlorine molecule (Cl_2) is covalent. Chlorine is therefore a simple covalent substance. Any covalent molecule has positive charges (nuclei) and negative charges (electrons), and the distribution of these charges is constantly changing. Occasionally, the charges will be arranged in such a way that the molecule forms a *temporary* or *instantaneous* dipole (that is, a molecule which has a positive end and a negative end). This dipole produces (or induces) other dipoles in neighbouring molecules, thus creating instantaneous-dipole-to-induced-dipole attractions, or van der Waals forces. The strength of these forces increases with the number of electrons in the molecule (or with the size of the molecule). For example, an iodine molecule is larger and has more electrons than a chlorine molecule. Therefore, the van der Waals forces between iodine molecules are stronger and can hold the molecules together as a solid under normal conditions. Note that while van der Waals forces in iodine are stronger than those in chlorine, they are generally very weak compared to covalent bonds.

It is important to distinguish between the covalent bonds *within* a molecule, called intra-molecular bonds, and the forces of attraction *between* individual molecules, called inter-molecular forces. For example, the evaporation of a simple covalent substance such as bromine involves breaking the weak van der Waals forces between the molecules, while the covalent bond within each molecule remains intact.

Since chlorine is a small molecule with weak van der Waals forces, it has a low boiling point and is therefore a gas under normal conditions.

Simple covalent substances have no free-moving charged particles, and hence they do not conduct electricity.

Ammonia, NH_3, reacts with hydrogen ions H^+ to produce the ammonium ion NH_4^+. In ammonia, the bonds between the nitrogen and hydrogen atoms are covalent. Name the type of bonding formed between the nitrogen atom and the hydrogen ion H^+ when NH_4^+ is formed, and explain how it is formed and how it differs from the covalent bonds in NH_3.

Answer

The bond is a co-ordinate bond.

The nitrogen atom in NH_3 has a lone pair of electrons, while the hydrogen ion H^+ is electron-deficient. The nitrogen atom uses its lone pair of electrons to form a co-ordinate bond with the hydrogen ion, thus forming NH_4^+.

In a covalent bond, each atom provides one of the two shared electrons, while in a co-ordinate bond the two shared electrons are donated by the same atom.

Background

A hydrogen atom contains one electron and one proton. When the hydrogen atom loses its single electron, it becomes the hydrogen ion H^+, which is simply a single proton. Since the positive charge of a proton is very highly concentrated, H^+ can easily attract and accept a pair of negative electrons from an atom that has a lone pair of electrons, such as O or N. In ammonia, NH_3, the nitrogen atom can share its lone pair of electrons with H^+, thus forming a new covalent bond. Since this covalent bond was formed by one atom donating both of the shared electrons, it is called a dative covalent bond, or a co-ordinate bond.

(i) Define the term *electronegativity*.

(ii) State and explain the trend in electronegativity across a period in the periodic table.

(iii) State and explain the trend in electronegativity down a group in the periodic table.

Answer

(i) Electronegativity is the relative ability of an atom to attract the electron density in a covalent bond.

(ii) Going across a period in the periodic table, electronegativity increases. This is because as we go across a period, the number of shells remains the same but the positive nuclear charge increases and the size of the atom decreases. This increases the ability of the atom to attract the shared electrons in a covalent bond.

(iii) Going down a group in the periodic table, electronegativity decreases because the size of the atom increases, so there is more shielding. The ability of the atom to attract electrons decreases.

Background

Electronegativity describes the ability of an atom to attract the bonding pair of electrons in a covalent bond. This ability to attract electrons depends on two factors:

- the number of protons in the nucleus (the nuclear charge), and
- the size of the atom

Atoms with more protons in the nucleus are more able to attract electrons. Also, smaller atoms are better at attracting electrons as the attracted electrons are then closer to the nucleus.

As we go across a period, the number of protons (i.e. the nuclear charge) increases. This means that the electrons in their shells will be pulled closer to the nucleus and so the atom becomes smaller. The higher nuclear charge and the smaller atomic size both result in an increase in the ability of the atom to attract the electrons in a covalent bond, i.e. an increase in electronegativity.

As we go down a group, the nuclear charge increases, which tends to increase electronegativity. At the same time, the atomic size increases (because there are more shells, so the outer electrons become increasingly more shielded, which tends to reduce electronegativity. Of these two factors, the effect of the increased size is more important, and therefore electronegativity decreases down the group.

Question 3.7

In hydrogen chloride HCl the bond between the chlorine atom and the hydrogen atom is a polar covalent bond.

    (i)   **State what is meant by a *polar covalent bond*.**

    (ii)   **Explain why the covalent bond in HCl is polar while that in chlorine Cl_2 is not.**

Answer

(i) Unequal sharing of the electrons in a covalent bond.

(ii) The H—Cl bond is polar because there is an electronegativity difference between chlorine and hydrogen. In Cl—Cl there is no difference in electronegativity between the two atoms and so the bond is not polar.

Background

When two atoms of different electronegativities form a covalent bond, the more electronegative atom will attract the shared pair of electrons closer to its nucleus, which means that its share of the electron density will be larger than that of the less electronegative atom. The electrons are not shared equally. The more electronegative atom becomes partially negative and the less electronegative atom becomes partially positive (since the positive charge in its nucleus now slightly exceeds the reduced negative charge around it). The covalent bond is then described as a polar bond. Since chlorine is more electronegative than hydrogen, the bond in HCl is polar ($H^{\delta+}$—$Cl^{\delta-}$). In Cl_2, the two atoms have the same electronegativity, and therefore the covalent bond is not polar.

Question 3.8

The following table shows the trend in relative electronegativity from carbon to fluorine in period 2 of the periodic table

Element	C	N	O	F
Electronegativity	2.5	3.0	3.5	4.0

(i)   Explain the trend shown in the table, and hence explain why fluorine is the most electronegative element in nature.

(ii) The relative electronegativity of hydrogen is 2.1. Arrange the following bonds in order of increasing polarity: H—N, H—C, H—F, H—O. Explain your answer.

Answer

(i) As we go across the period from C to F, the number of shells remains the same but the nuclear charge increases and the atomic size decreases. Therefore, the ability of the atom to attract electrons increases.

Fluorine is the most electronegative element in nature because its atom has the highest charge to size ratio among all the elements, since:

(1) it is the smallest atom in period 2

(2) it has the highest nuclear charge in period 2

(ii) H—C < H—N < H—O < H—F

Electronegativity increases from C to N to O to F. Therefore, the difference in electronegativity increases from H—C to H—N to H—O to H—F. Polarity increases as the difference in electronegativity between the two bonded atoms increases.

Question 3.9

The table below shows the boiling points of the hydrogen halides

Hydrogen halide	HF	HCl	HBr	HI
Boiling point in K	293	188	206	238

    (i)   Name the strongest type of intermolecular forces in HF and explain how it arises.

    (ii)   Explain the trend in the boiling point of the hydrogen halides shown in the table.

    (iii)   Water has a boiling point of 373 K. Suggest a reason why water has a higher boiling point than hydrogen fluoride.

Answer

(i) Hydrogen bonding.

Fluorine is a highly electronegative atom. In HF, the fluorine atom attracts the electron density in the H—F bond, making the hydrogen atom electron-deficient and therefore partially positive. The partially positive hydrogen atom will then be attracted by electrostatic forces to one of the lone pairs of electrons on the fluorine of a neighbouring HF molecule, thus forming a hydrogen bond.

(ii) HF has the highest boiling point amongst the hydrogen halides because of the strong hydrogen bonds between its molecules. The boiling point increases from HCl to HBr to HI because of the increasing size of the molecule. As the size of the molecule increases, the Van der Waals forces between the molecules become stronger, and more energy is needed to overcome them.

(iii) A water molecule contains two H atoms, while a hydrogen fluoride molecule contains only one. Therefore, H_2O can potentially form more hydrogen bonds than HF. More energy will be needed to break the larger number of hydrogen bonds in water.

Background

When a hydrogen atom is covalently bonded to a highly electronegative atom (F, O, or N), the electron density of the covalent bond is shifted from the hydrogen atom toward the more electronegative atom. The hydrogen atom becomes partially positive while the electronegative atom becomes partially negative. This results in an attractive force between the hydrogen of one molecule and the electronegative atom of another, as shown in Figure 3.3. This force is the hydrogen bond (represented by dashed lines in Figure 3.3).

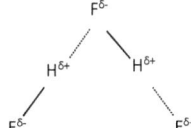

Figure 3.3: Hydrogen bonding in HF

Fluorine, oxygen and nitrogen are the only elements capable of causing a sufficiently high partial charge on the hydrogen atom to result in hydrogen bonding.

The hydrogen bond is relatively strong and results in a high boiling point. This is why HF has the highest boiling point of all the hydrogen halides. HCl, HBr, and HI, on the other hand, form weak van der Waals forces, which are much weaker than the hydrogen bond, and therefore the boiling points of these halides are lower than that of HF. As the size of the halogen atom increases from Cl to Br to I, the strength of the van der Waals forces increases from HCl to HBr to HI (see Question 3.4), and so the boiling point increases.

Both water and hydrogen fluoride have hydrogen bonds. In water, there are two hydrogen atoms and two lone pairs of electrons on the oxygen. This means that there are exactly the right numbers of H atoms and lone pairs so that every one of them is involved in hydrogen bonding. In hydrogen fluoride, there are three lone pairs on the fluorine but only one hydrogen atom, so there is a shortage of hydrogen atoms. As a result, there are more hydrogen bonds in water than in hydrogen fluoride.

Question 3.10

The following graph shows the trend in the boiling points of three halogen halides, HI, HBr, HCl.

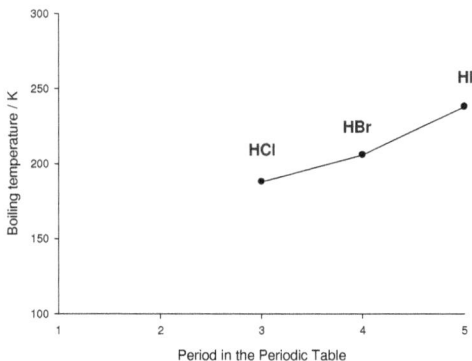

(i)   Explain this trend in terms of intermolecular forces.

(ii)   Assuming that the fourth hydrogen halide HF followed this trend, indicate on the graph where approximately you would expect its boiling point to be.

(iii)   The actual boiling point of HF is 293 K. Explain why this is very different from the boiling point HF would have based on the trend shown.

Answer

(i) HCl, HBr and HI have weak van der Waals forces. The strength of the van der Waals forces increases as the size of the molecule increases from HCl to HBr to HI, and so the boiling point increases in the same order.

(ii)

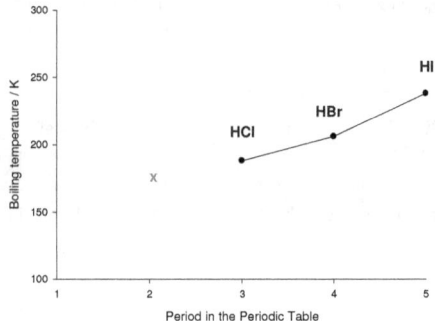

(iii) Among all the hydrogen halides, HF is the only one to contain hydrogen bonds. The hydrogen bond is stronger than van der Waals forces of attraction, and therefore HF has a much higher boiling point than would be expected from the trend shown.

Background

If HF had only van der Waals forces between its molecules, its boiling point would be lower than that of HCl, since HF is a smaller molecule. However, HF has strong hydrogen bonds, which explains why it has a boiling point higher than all the hydrogen halides (see Question 3.9).

Question 3.11

Ammonia and methane have similar relative molecular masses. The boiling point of ammonia is 239.5 K, while the boiling point of methane is 109 K. Explain why the boiling point of ammonia is much higher than that of methane.

Answer

The strongest type of intermolecular forces in ammonia is hydrogen bonding, while the intermolecular forces in methane are van der Waals forces. Hydrogen bonds are much stronger than van der Waals forces and need much more energy to overcome.

Background

The strength of the van der Waals forces depends largely on the relative molecular mass (M_r) of the molecule. Since ammonia and methane have similar M_r values, they have more or less the same van der Waals forces. However, the intermolecular forces in ammonia also include hydrogen bonding, which is much stronger than the van der Waals forces in methane. The boiling point of ammonia is therefore much higher than that of methane.

Question 3.12

Both water (H_2O) and ammonia (NH_3) form hydrogen bonds. Explain why the boiling point of water (373 K) is much higher than that of ammonia (239.5 K).

Answer

In water, the oxygen atom has two lone pairs of electrons, while in ammonia, the nitrogen has only one lone pair of electrons. This means that there are more hydrogen bonds in water compared with ammonia, and therefore more energy is needed to break them.

Question 3.13

Under normal conditions of temperature and pressure, chlorine is a gas, bromine is a liquid, and iodine is a solid. Explain this in terms of intermolecular forces.

Answer

Chlorine, bromine, and iodine are simple covalent molecules. The strength of the intermolecular forces of attraction increases with M_r, which in turn increases from chlorine to bromine to iodine. Chlorine has weak van der Waals forces and it's therefore a gas, bromine has stronger van der Waals forces and it's therefore a liquid, while iodine has even stronger van der Waals forces and it's therefore a solid.

Question 3.14

State and explain the trend in the boiling points of the noble gases from He to Xe.

Answer

The boiling point of the noble gases increases from He to Xe. This is because the strength of the van der Waals forces between the atoms increases as the relative atomic mass increases from He to Xe, and so more energy is needed to overcome these forces.

Question 3.15

For each of the following molecules, name the shape of the molecule and state whether or not it is a permanent dipole: CO_2, CH_4, CCl_2F_2, NH_3, H_2O.

Answer

CO_2: linear – not a dipole

CH_4: tetrahedral – not a dipole

CCl_2F_2: tetrahedral – dipole

NH_3: trigonal pyramidal – dipole

H_2O: V-shaped (or bent) – dipole

Background

The electron pair repulsion theory is used to determine the shape of a molecule. According to this theory, the shape of a covalent molecule is determined by the positions of the electron pairs surrounding the central atom. Any two pairs of electrons around an atom will repel each other and will take up positions as far apart as possible to minimise this mutual repulsion. The shape of the molecule will be such that the angle between the pairs of electrons is maximised and the repulsion minimised. A pair of electrons around an atom can be a shared pair (a covalent bond) or a lone pair. Lone pairs of electrons have a more concentrated negative charge and will therefore repel more strongly than bonding pairs, so repulsion increases in the order:

bonding pair-bonding pair $<$ bonding pair-lone pair $<$ lone pair-lone pair

When two atoms of different electronegativities are covalently bonded, the more electronegative atom attains a partial negative charge ($\delta -$) while the other atom attains a partial positive charge ($\delta +$). The bond is said to be polar. The polarity of a molecule is determined by the sum of the effects of the polarities of all the bonds in the molecule. If the positive charge is concentrated toward one end of the molecule and the negative charge is concentrated toward the opposite end, the molecule is polar (a dipole).

CO_2: the central atom, C, has 4 electrons in its outer shell and forms two double bonds with the two oxygen atoms. All the electrons in the outer shell of C are involved in bonding, and so C has no lone pairs of electrons. The atoms are therefore arranged in a linear molecule with a bond angle of 180°.

Each double bond in CO_2 is polar because oxygen is more electronegative than carbon. However, due to the linear shape of the molecule, the effects of the two polar bonds cancel out, and the molecule is not a dipole.

CH_4: The carbon atom has 4 electrons in its outer shell and forms four single covalent bonds with the four hydrogen atoms in methane. The four bonds are arranged in a tetrahedron that has a bond angle of 109.5°.

While each of the C—H bonds is polar, the shape of the molecule means that the polarities of the four bonds cancel out and therefore the molecule is not a dipole.

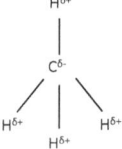

CCl_2F_2: This is similar to CH_4. The molecule is tetrahedral with a bond angle of 109.5°.

Both fluorine and chlorine are more electronegative than carbon, and so the bonds $C^{\delta+}$—$F^{\delta-}$ and $C^{\delta+}$—$Cl^{\delta-}$ are polar. However, due to the shape of the molecule, the effects of the four polar bonds cancel out and the molecule is not a dipole.

NH_3: The central atom, N, has 5 electrons in its outer shell, three of which are used to make covalent bonds with the hydrogen atoms, while the remaining two electrons form a lone pair. The four electron pairs (three bonds and one lone pair) form a tetrahedron, but this tetrahedron is distorted because one of the electron pairs is a lone pair. Since lone pairs repel more than bonding pairs, the increased repulsion due to the lone pair will squeeze the hydrogen atoms closer together thus reducing the angle between the bonds from 109.5° to 107°. The molecule, including the lone pair, is a distorted tetrahedron, but the atoms on their own form a triangular pyramid.

Since nitrogen is more electronegative than hydrogen, the three N—H bonds are polar. The effects of the polarity of the three bonds do not cancel, and so the molecule is a dipole.

H_2O: The oxygen atom has six electrons in the outer shell. Two electrons are used to make two O—H bonds, and the remaining four electrons form two lone pairs. Therefore, there are four pairs of electrons arranged in a distorted tetrahedron. The two bonds, however, form a V-shape. Since there are two lone pairs repelling the O—H bonds, the bond angle is 104.5°, smaller than that in NH_3.

Oxygen is more electronegative than hydrogen and the two O—H bonds are therefore polar. As H_2O is a V-shaped molecule, the effects of the polarity of the two bonds do not cancel, and the molecule is therefore a dipole.

Question 3.16

(i) Name the shape of the molecule BF_3 and explain why each bond angle is exactly 120°.

(ii) Predict the shape of the molecule CCl_2 stating the bond angle. Explain why this angle is different from that in BF_3.

Answer

The molecule of BF_3 is trigonal planar. The bond angle in a trigonal planar molecule is 120°. The angle is exactly 120° because the three pairs of electrons around the central atom are bonding pairs and so they repel each other equally.

(ii) CCl_2 is a bent (or V-shaped) molecule with a bond angle of around 118°.

The bond angle is different from that in BF_3 because carbon in CCl_2 has a lone pair of electrons, and lone pairs of electrons repel more than bonding pairs.

Question 3.17

The methane molecule CH_4 has four covalent bonds arranged in a tetrahedral shape. There are also four covalent bonds between chlorine and fluorine in the ion ClF_4^-.

(i) State the bond angle in CH_4.

(ii) Deduce the total number of electron pairs (lone and bonding) surrounding the central chlorine atom in ClF_4^-.

(iii) State and explain the position of any lone pairs of electrons in ClF_4^-.

(iv) Hence deduce the shape of ClF_4^- stating the bond angle.

Answer

(i) 109.5°.

(ii) Six pairs of electrons.

(iii) The two lone pairs of electrons are above and below the plane formed by the four bonds in order to minimise the mutual repulsion.

(iv) Square planar with a bond angle of 90°.

Background

A neutral chlorine atom has seven electrons in its outer shell. In ClF_4^- there is a 1^- charge on the ion, which means that the chlorine atom has an extra electron. This makes the total number of electrons around the chlorine atom 8. Four of these eight electrons are used to make four covalent bonds by pairing up with four electrons from the fluorine atoms, leaving two lone pairs of electrons. So, we have a total of six pairs of electrons. The four bonding pairs make a square planar shape, while the two lone pairs are positioned one above and one below this plane. This places the lone pairs as far apart as possible in order to minimise the repulsion between them (since the repulsion between lone pairs is the strongest).

Question 3.18

Explain how the electron pair repulsion theory can be used to determine the shape of NCl_3, and hence deduce the shape of this molecule, stating the bond angle.

Answer

The central atom in this molecule, N, is surrounded by three shared pairs of electrons and one lone pair. The four pairs will be positioned as far apart as possible, which leads to a shape based on a tetrahedron. The three bonds are arranged in a triangular pyramid. Lone pairs repel more than bonding pairs, and so the bond angle is 107°.

Background

The central atom in NCl_3 is N with five electrons in its outer shell. Three of the five electrons are used to form covalent bonds with the chlorine atoms, leaving two electrons as a lone pair. This means that the N atom is surrounded by four pairs of electrons in total, and the shape is therefore based on a tetrahedron. The angle in a regular tetrahedron is 109.5°. However, one of the pairs around N is a lone pair, and since lone pairs repel more than bonding pairs, the bonds in NCl_3 will be pushed closer together by the extra repulsion of the lone pair, thus reducing the bond angle to 107°. Note that while the arrangement of the four pairs of electrons forms a distorted tetrahedron, the three bonds form a triangular pyramid. These are two different ways of looking at the shape of the molecule.

Question 3.19

Use the electron pair repulsion theory to predict the shape of, and the bond angle in, $BeCl_4^{2-}$. Briefly explain how you arrived at your answer.

Answer

In $BeCl_4^{2-}$, the Be atom is surrounded by four covalent bonds and has no lone pairs of electrons, so it has a tetrahedral shape with a bond angle of 109.5°.

Background

A neutral beryllium atom has two electrons in the outer shell, but in the $BeCl_4^{2-}$ ion it has two more electrons, making the total four. All of the four electrons are involved in covalent bonding with chlorine, so the shape is a regular tetrahedron.

Question 3.20

The following table compares the properties of four different types of substance. Use your knowledge of bonding and structure to complete the table.

Melting point	Conductivity in the solid state	Conductivity in the liquid state	Type of structure	Example
High	Yes	Yes		
			Giant ionic	
Very low				Sulphur
	No	No		Diamond

Answer

Melting point	Conductivity in the solid state	Conductivity in the liquid state	Type of structure	Example
High	Yes	Yes	Metallic	Magnesium
High	No	Yes	Giant ionic	Sodium chloride
Very low	No	No	Simple covalent	Sulphur
Very high	No	No	Giant covalent	Diamond

Background

Metallic structure:

Metals have high melting points because they have giant structures. The metallic structure is composed of positive ions surrounded by a sea of delocalised (free-moving) electrons. The attraction between the ions and the delocalised electrons in the giant structure is strong and requires a large amount of energy to overcome.

Metals are good conductors of electricity in the solid and liquid states because they have free-moving electrons which can carry the electric current.

Giant ionic structure

Ionic substances have high melting points as they have giant structures. In the ionic lattice, each ion is attracted to several oppositely charged ions. It requires a large amount of energy to break the large number of ionic bonds in the lattice.

Ionic compounds do not conduct electricity in the solid state because the ions are not free to move. In the liquid state, however, the ions are free to move, and so ionic substances are good conductors in the liquid state.

Giant covalent structure

Diamond has a giant covalent structure in which each carbon atom is covalently bonded to four other carbon atoms forming a giant structure. A very large amount of energy is needed to break the large number of very strong covalent bonds in diamond. This explains why giant covalent substances have very high melting points.

Giant covalent substances are usually poor conductors of electricity as they have no free-moving electrons. Graphite, however, is an exception. In its giant structure, graphite has delocalised electrons since each carbon atom forms three covalent bonds, leaving one electron unshared and therefore free to move.

Simple molecular structures

Simple covalent molecules are attracted to each other by weak intermolecular forces such as van der Waals forces. Such substances usually have low melting points since only a small amount of energy is required to overcome the weak intermolecular forces between the molecules. Since there are no charged particles to carry charge, simple molecular substances do not conduct electricity.

Question 3.21

Iodine and chlorine are examples of the simple molecular structure. Explain in terms of intermolecular forces why iodine is a solid at room temperature while chlorine is a gas.

Answer

Iodine molecules have a larger number of electrons than chlorine molecules. Therefore, the van der Waals forces in iodine are strong enough to hold the molecules together as a solid. In chlorine, the van der Waals forces are much weaker and cannot hold the molecules close together.

Question 3.22

Iodine has a simple molecular structure. By reference to this structure, explain the following properties of iodine:

(a) Iodine crystals are soft
(b) Iodine crystals sublime easily to form gaseous iodine
(c) Iodine does not conduct electricity

Answer

(a) The van der Waals forces in the iodine crystal are very weak compared to covalent bonds, and therefore the iodine crystal is soft and breaks easily.

(b) Iodine sublimes easily because a small amount of energy is needed to break the weak van der Waals forces in the iodine crystal.

(c) There are no charged particles in the structure of iodine to carry charge.

Question 3.23

The melting point of phosphorous (P_4) is 317 K, while the melting point of sulphur (S_8) is 386 K. Explain this difference.

Answer

Both phosphorous and sulphur have a simple molecular structure. The molecule of sulphur (S_8) has more electrons than the molecule of phosphorous (P_4), and therefore the van der Waals forces between the sulphur molecules are greater and the melting point is higher.

Diamond has a giant covalent (macromolecular) structure.

(i)   State the bond angle in diamond.
(ii)   Explain how the structure of diamond gives rise to the following properties:
        (a) Diamond is a very hard material
        (b) Diamond has a very high melting point of around 3820 K
        (c) Diamond does not conduct electricity

Answer

(i) 109.5°.

(ii)   (a) The carbon atoms in diamond form a giant three-dimensional lattice with a large number of very strong covalent bonds.

(b) A large amount of energy is needed to break the large number of strong covalent bonds in diamond.

(c) There are no free charged particles in diamond to carry the electric current.

Graphite and diamond are both forms of carbon.

(i)   Graphite and diamond are examples of the same type of structure. Name this type of structure.
(ii)   State the bond angles in diamond and graphite.
(iii)   Explain the features of the structure of graphite that allow it to be used as a lubricant and in pencils.
(iv)   Compare the electrical conductivities of graphite and diamond.

Answer

(i) Macromolecular structure.

(ii) Graphite: 120°. Diamond: 109.5°.

(iii) Graphite has a layered structure. The layers are held together by weak van der Waals forces. This means that the layers can slide across one another easily, making graphite soft and flaky. This softness makes graphite suitable as a lubricant, and the flakiness allows the graphite to transfer from the pencil to the paper.

(iv) Diamond does not conduct electricity as it has no charged particles. Graphite has delocalised electrons which allow it to conduct electricity along its layers.

Question 3.26

Describe the structure of graphite and explain how it gives rise to the following observation

    (a) Graphite has a high melting point
    (b) Graphite is used as a lubricant
    (c) Although it is a non-metal, graphite conducts electricity

Answer

Graphite has a macromolecular structure. Each carbon atom in graphite forms three covalent bonds with three neighbouring carbon atoms, leaving one electron free. The carbon atoms form layers of hexagons that are held together by weak van der Waals forces.

(a) Graphite has a very high melting point because a large amount of energy is needed to break the large number of strong covalent bonds in its giant structure.

(b) Graphite is used as a lubricant because its layers are held together by weak van der Waals forces and can easily slide across one another.

(c) Graphite conducts electricity because it contains delocalised electrons in its structure.

Question 3.27

Although carbon is a non-metal, one of its forms, graphite, conducts electricity. Explain how graphite conducts differently from metals.

Answer

In graphite, the delocalised electrons are spread along the layers of carbon atoms, thus allowing graphite to conduct electricity well along the layers but poorly between the layers. Metals, on the other hand, conduct well in all directions as the sea of delocalised electrons is spread throughout the whole structure.

Question 3.28

(i) Explain how the ions in sodium metal are held together.

(ii) Explain why metals are malleable.

(iii) Which do you expect to have a higher melting point, sodium or aluminium? Explain your answer.

Answer

(i) The sodium ions are held together by electrostatic attraction to the sea of delocalised electrons surrounding the ions.

(ii) Metals have a layered structure. When a force is applied, the layers of ions in the lattice slide over each other without the lattice breaking up.

(iii) Aluminium will have a higher melting point than sodium. The aluminium ion is smaller and has a higher charge than the sodium ion. Also, there are more delocalised electrons in aluminium. Therefore, the metallic bond in aluminium is stronger and requires more energy to overcome.

Background

(i) Although positive ions tend to repel each other, the positive ions in a metal lattice are held together by their attraction to the sea of delocalised electrons that surrounds them.

(ii) Metals are malleable, which means that they can be shaped into different shapes. This is because the ions in a metal lattice are arranged in layers. When a force is applied, the layers can slide over each other without causing the structure to break up or shatter.

(iii) The strength of the metallic bond depends on the charge and size of the ions and the number of delocalised electrons. Sodium and aluminium both have the same number of electron shells. However, aluminium has three electrons in the outer shell and sodium has only one. Therefore, aluminium forms an ion with a 3+ charge (Al^{3+}), while sodium forms an ion with a 1+ charge (Na^+). This means that Al^{3+} is more positive than Na^+ and is also smaller in size. It also means that there are more delocalised electrons in aluminium because each atom loses 3 electrons whereas each sodium atom loses only one electron. This means that the metallic bond in aluminium is stronger and so more energy is needed to cause aluminium to melt.

4. PERIODICITY

Question 4.1

The area of the periodic table comprising groups 1 and 2 is called the s-block. Explain why there are only two groups in this block.

Answer

The elements in the s-block have their outer electrons in the s-orbital, which can accommodate up to two electrons.

Background

If an atom has its outermost electrons in the s-orbital, then it is placed in the s-block of the periodic table. The group in which the atom is placed depends on the number of electrons in the outer shell of the atom. Group 1 elements have one electron in their outer shell (in the s-orbital). Group 2 elements have two electrons in the outer shell (also in the s-orbital). Since the s-orbital can accommodate up to two electrons, this means that the s-block is made up of two groups only. Elements with three electrons in the outer shell have their outermost electron in a p-orbital, therefore group 3 elements are part of the p-block.

Question 4.2

Explain why period 1 of the periodic table contains only two elements, while period 2 contains eight elements.

Answer

The elements of period 1 have only one energy level, while the elements of period 2 have two energy levels. The first energy level accommodates a maximum of two electrons, while the second energy level accommodates a maximum of eight electrons.

Background

In the periodic table, atoms are placed in periods depending on the number of energy levels (or shells) in the atom. So, an atom with only one energy level will be found in period 1, an atom with two energy levels will be in period 2, and so on. The first energy level accommodates up to two electrons, and so period 1 will contain two elements: one with only one electron, and the other with two electrons. The second energy level accommodates up to eight electrons, and so period 2 contains eight elements.

Question 4.3

All the elements which have their outer shell electrons in the p-orbitals are placed in the p-block. The electron arrangement of helium He is $1s^2$. Explain why helium is placed in the p-block.

Answer

Helium has the properties of noble gases, and so it is placed with the noble gases in group 8, which is in the p-block.

Background

The elements that make up a group (a column) in the periodic table have similar properties. For example, fluorine, chlorine, bromine and iodine (the halogens) have similar chemical and physical properties and they are found in the same group in the periodic table. They also have their outer electrons in the p-orbitals; therefore their group is found in the p-block. Helium has only two electrons, and these electrons occupy an s-orbital, so you would expect to see helium in the s-block. However, since helium has the properties of noble gases (e.g. unreactive, very low melting and boiling points, etc.), it belongs to the group of noble gases (group 8 or 0), although this group is in the p-block (since all the noble gases, apart from helium, have their outer electrons in the p-orbitals).

Question 4.4

The figure below shows the melting points of period 3 elements. Use your knowledge of bonding and structure to explain the trend shown.

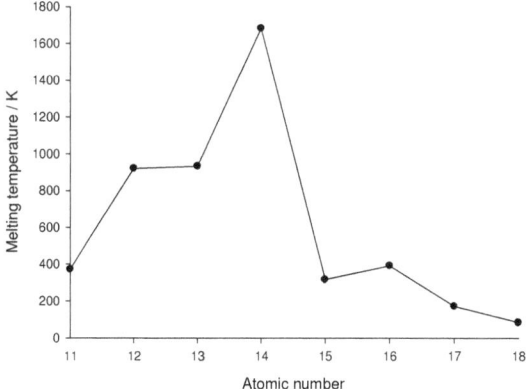

Answer

The melting points of Na, Mg, and Al are relatively high due to the metallic bond. The strength of the metallic bond increases from Na to Al as the charge on the metal ion increases, the number of delocalised electrons increases, and the size of the ion decreases. Therefore, the melting point increases from Na to Al.

Silicon has a very strong giant covalent structure. A large amount of energy is needed to break the large number of strong covalent bonds, and therefore silicon has the highest melting point in period 3.

The elements P, S, and Cl have simple molecular structures, with weak van der Waals forces between the molecules, and so they have low melting points. The number of electrons in the molecule decreases in the order $S_8 > P_4 > Cl_2$ and therefore the melting point decreases in the same order. Argon has a simple atomic structure with even weaker van der Waals forces, and therefore it has the lowest melting point.

Background

Elements in the same period show large variations in their physical properties. For example, the elements in period 3 have different structures, and therefore they have widely different melting points. Na, Mg and Al are metals and have metallic structures. The strength of the metallic bond increases as (i) the charge on the metal ion and the number of delocalised electrons increase, and (ii) the size of the ion decreases. Each Na atom contributes one electron to the delocalised sea of electrons, forming the ion Na^+; Mg loses two electrons, thus forming Mg^{2+}; while Al loses three electrons, forming the ion Al^{3+}. Therefore, the strength of the metallic bond increases in the order Na < Mg < Al, and so the melting point increases in the same order.

Silicon has a macromolecular (giant covalent) structure. The silicon atoms are held together by very strong covalent bonds in a large 3D lattice, making it very difficult to separate the atoms and cause silicon to melt.

Phosphorus, sulphur and chlorine have simple molecular structures in which the molecules are held together by very weak van der Waals forces. However, as the number of electrons in the molecule increases, the strength of the van der Waals forces increases. Phosphorus exists as P_4, sulphur as S_8 and chlorine as Cl_2. S_8 is larger than P_4 and Cl_2 and has the highest number of electrons, while Cl_2 is the smallest and has the lowest number of electrons. The strength of the van der Waals forces decreases in the order $S_8 > P_4 > Cl_2$, and so the melting point decreases in the same order.

Argon has a simple atomic structure (it exists as individual atoms), with fewer electrons than a chlorine molecule Cl_2, therefore it has weaker van der Waals forces and hence a lower melting point.

State and explain the trends in atomic radius

    (i)   across a period

    (ii)   down a group

Answer

(i)   The atomic radius decrease across a period. The atoms in the same period have the same number of shells, but as we move across a period, the nuclear charge increases. This increased charge pulls the electrons in closer to the nucleus, and therefore the size of the atom decreases.

(ii) Going down a group, the atomic radius increases because the number of electron shells increases.

Background

The size of an atom (or the atomic radius) depends on two factors:

    i.   The number of shells.

    ii.   The nuclear charge.

For elements with the same number of shells (i.e. elements in the same period), the nuclear charge increases across the period, and so the attraction between the nucleus and the electrons also increases. The increased nuclear charge pulls the electrons in closer to the nucleus, making the atom smaller in size.

On the other hand, as we go down a group, the number of energy levels or shells surrounding the nucleus increases, and therefore the size of the atom increases.

The values of the first ionisation energies of period 3 elements show that:

    (i)   There is a general increase in ionisation energy across the period.

    (ii)   There is a drop in ionisation energy from magnesium (group 2) to aluminium (group 3).

    (iii)   There is a drop in ionisation energy from phosphorus (group 5) to sulphur (group 6).

Explain the above observations.

Answer

(i) As we go across a period, the number of shells remains the same but the nuclear charge increases, making it increasingly difficult to remove an electron from the outer shell.

(ii) In magnesium, the first electron to be removed is in a 3s orbital, while in aluminium, the first electron to be removed is in a 3p orbital. The 3p orbital is further away from the nucleus, and so electrons in 3p are easier to remove.

(iii) The first electron to be removed from phosphorus is an unpaired electron, while the first electron to be removed from sulphur is a paired electron. The paired electron is easier to remove because it is already being repelled by another electron so it requires less energy to remove from the atom.

Background

The amount of energy needed to remove an electron from a gaseous atom depends on the strength of the attraction between the nucleus and that electron. This attraction becomes stronger as:

(i)   The nuclear charge (the number of protons in the nucleus) increases

(ii)   The atomic radius decreases

The electrons, being negative, are attracted to the positive protons in the nucleus. The more protons the nucleus contains, the stronger the attraction. Also, the closer the electrons are to the nucleus, the less they are shielded from it, and the stronger the attraction.

As we go across a period from left to right, the number of shells remains the same but the nuclear charge increases, causing the electrons to be drawn closer to the nucleus, thus reducing the size of the atom. The increased nuclear charge and the smaller atomic size increase the attraction between the nucleus and the electrons, making it increasingly more difficult to remove an electron from the outer shell, and so ionisation energy increases. However, there are two breaks in this pattern:

i.   Ionisation energy decreases from group 2 to group 3 (from Mg to Al in period 3). Group 2 elements have their outer electron in an s-orbital, while group 3 elements have their outer electron in a p-orbital. A p-orbital is further away from the nucleus than an s-orbital in the same shell, and so it is easier to ionise.

ii.   Ionisation energy also decreases from group 5 to group 6. For a group 5 element, the electron to be removed is a single (unpaired) electron; while for a group 6 element, the electron is paired and is already being repelled by another electron. Removing a paired electron requires less energy than removing an unpaired electron.

As we go down a group, the number of shells increases, so the outer electrons become increasingly more shielded from the nucleus, making them easier to remove from the atom, so ionisation energy decreases.

Question 4.7

The successive ionisation energies of element X are given below (in kJ/mol):

1681	3374	6050	8408	11023	15164	17868	92038	106434

Determine the position (group number and period number) of element X in the periodic table, and hence deduce its identity. Explain your answer.

Answer

There is a gradual increase in ionisation energy from the 1st to the 7th value. This means that element X has seven electrons in its outer shell and is therefore in group 7. There is a very large increase from the 7th value to the 8th value, indicating that the 8th electron is in a new shell. The atom therefore has two shells and so it belongs to period 2. The element in group 7 and period 2 is fluorine.

Background

If we look closely at the values in the table above (also plotted in Figure 4.1), we see that the ionisation energy increases gradually from the 1st to the 7th value. This means that the outer shell contains 7 electrons, the first requiring 1681 kJ/mol to remove from the atom, the second 3374 kJ/mol, and so on. We deduce that the element must be in group 7 (the group number equals the number of electrons in the outer shell). Then, from the 7th to the 8th value there is a very large increase in ionisation energy, which means that the 8th electron must be in an inner shell that is much closer to the nucleus compared to the 7th electron. So the atom has 2 shells and is therefore in period 2 (the period number is equal to the total number of shells). The element in period 2 and group 7 is fluorine.

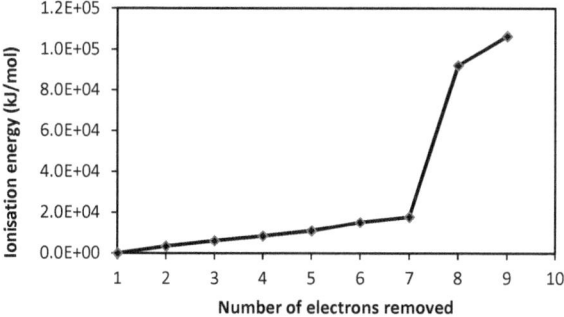

Figure 4.1: Successive ionisation energies of element X.

Question 4.8

The first ionisation energies of magnesium and aluminium are given below.

	Atomic number	First ionisation energy (kJ/mol)
Mg	12	738
Al	13	578

Explain why, although aluminium has a higher atomic number than magnesium, it has a lower first ionisation energy.

Answer

The electron to be removed from aluminium is in a p-orbital, while the electron to be removed from magnesium is in an s-orbital. An electron in the p-orbital is further away from the nucleus compared to an electron in the s-orbital of the same shell, and so is easier to remove.

Question 4.9

Sodium, Na, is a metal placed in group 1 of the periodic table. The first ionisation energy of sodium is 496 kJ/mol. Another metal, X, in the same group has a first ionisation energy of 520 kJ/mol. Deduce the identity of X and explain your answer.

Answer

The metal X is lithium Li. Ionisation energy decreases as we go down a group. Since the first ionisation energy of X is higher than that of Na, then X must be above Na in the periodic table. The only metal above Na is Li.

5. INTRODUCTION TO ORGANIC CHEMISTRY

Question 5.1

Fill in the missing information in the following table to give the name, structural formula, and displayed formula of the compounds shown.

Name of compound	Structural formula	Displayed formula
		$H_3C-CH_2-CH(CH_3)-CH_2-CH_3$ chain: H—C—C—C—C—C—H with H's above and below, and H—C—H branch below third carbon
	$CH_3CH_2CBr(CH_3)_2$	
2-methylpent-2-ene		
		H—C—C—C—C—C—H chain with top: H H H H Cl; bottom: H H Br Cl H
1-bromo-2-iodopropane		
		F—C—Cl with Cl above and F below
2,3-dimethylbutane		
		O—C—C—C—H with top: H H Br; bottom H (on O), H H H

Answer

Name	Structural formula	Displayed formula
3-methylpentane	$CH_3CH_2CH(CH_3)CH_2CH_3$	H atoms on a 5-carbon chain with a CH_3 branch on the central carbon
2-bromo-2-methylbutane	$CH_3CH_2CBr(CH_3)_2$	4-carbon chain with Br and CH_3 branch on second carbon
2-methylpent-2-ene	$CH_3CH_2CHC(CH_3)_2$	chain with C=C double bond and CH_3 branch
3-bromo-1,2,-dichloropentane	$CH_3CH_2CHBrCHClCH_2Cl$	5-carbon chain with Cl, Br, Cl substituents
1-bromo-2-iodopropane	$CH_2BrCHICH_3$	3-carbon chain with Br and I substituents
dichlorodifluoromethane	CCl_2F_2	single carbon with 2 Cl and 2 F
2,2-dimethylbutane	$CH_3CH_2C(CH_3)_2CH_3$	4-carbon chain with two CH_3 branches on second carbon
3-bromopropan-1-ol	$CH_2BrCH_2CH_2OH$	3-carbon chain with OH and Br substituents

Background

To name simple organic molecules (alkanes and alkenes), follow the steps below:

1. Name the longest unbranched carbon chain based on the number of carbon atoms in the chain. If there are no C=C double bonds, use the name of the corresponding alkane. The table below gives the names of the first six alkanes.

Number of	Alkanes		Alkenes		Alkyl groups	
carbon atoms	Name	Formula	Name	Formula	Name	Formula
1	Methane	CH_4	-	-	Methyl	$-CH_3$
2	Ethane	C_2H_6	Ethene	C_2H_4	Ethyl	$-C_2H_5$
3	Propane	C_3H_8	Propene	C_3H_6	Propyl	$-C_3H_7$
4	Butane	C_4H_{10}	Butene	C_4H_8	Butyl	$-C_4H_9$
5	Pentane	C_5H_{12}	Pentene	C_5H_{10}	Pentyl	$-C_5H_{11}$
6	Hexane	C_6H_{14}	Hexene	C_6H_{12}	Hexyl	$-C_6H_{13}$

Alkenes contain at least one C=C double bond. To name an alkene, we use the suffix *–ene* instead of the suffix *–ane* at the end of the name of the alkane with the same number of carbon atoms, as shown in the table above.

2. Add a prefix to name any side carbon chain. A side chain is called an alkyl group and is named from the corresponding alkane with the same number of carbon atoms, as shown in the table above. The line '—' in front of the alkyl group means that the alkyl group is bonded to another atom. Note that an alkyl group is derived from the corresponding alkane by removing one hydrogen atom.

3. Indicate the position of the side chains. To do this, we number the carbon atoms in the longest unbranched chain starting from the side that is nearest to the alkyl group. For example, consider the following molecule

There are five carbon atoms in the longest unbranched carbon chain. The branching alkyl group (—CH₃) is closer to the right end of the molecule than to the left end. So, we number the carbon atoms starting from the right, which means that the side chain is on the second carbon. The side chain (—CH₃) is a methyl group, and the longest unbranched carbon chain contains five carbon atoms, corresponding to pentane. Therefore, the name of the above molecule is 2-methypentane. Now, try naming this molecule

$$H_3C-CH\overline{}CH_3$$
$$|$$
$$CH_2$$
$$|$$
$$CH_3$$

Here, the longest carbon chain contains four carbon atoms, corresponding to butane. They can be numbered as follows

$$\overset{2}{H_3C}-\overset{}{CH}\overline{}\overset{1}{CH_3}$$
$$|$$
$$_3\,CH_2$$
$$|$$
$$_4\,CH_3$$

The unnumbered —CH_3 group is a methyl side group attached to the second carbon. So the name is 2-methylbutane.

Note that if there are two of the same alkyl group, we add *di* to the beginning of the alkyl group. So, the molecule

$$CH_3$$
$$|$$
$$H_3C-C-CH_3$$
$$|$$
$$CH_3$$

is 2,2-dimethypropane.

4. If the compound contains a halogen atom (F, Cl, Br, I), a 'halo-prefix' (fluoro-, chloro-, bromo-, iodo-) is used, and the position of the halogen atom is indicated. If there are two different halogen atoms, they are written in alphabetical order. So,

$$\begin{array}{ccccc} H & H & H & H & Cl \\ | & | & | & | & | \\ H-C & -C & -C & -C & -C-H \\ | & | & | & | & | \\ H & H & Br & Cl & H \end{array}$$

is named 3-bromo,1-2-dichloropentane. Again, numbering the carbon atoms starts from the closest end to any side chain or functional group. A functional group is any reactive group of atoms added to the chain of carbon atoms, such as —OH (an alcohol group). Most functional groups are named by using a suffix. For example,

$$\begin{array}{ccc} H & H & H \\ | & | & | & \quad H \\ H-C-C-C-O & \nearrow \\ | & | & | \\ H & H & H \end{array}$$

is propan-1-ol. When there is a halogen atom attached to an alcohol, the alcohol functional group has priority and so we count from the nearest end to the —OH group. The molecule

$$
\begin{array}{cccc}
& H & H & Br \\
& |_1 & |_2 & |_3 \\
O\!-\!\!& C\!-\!\!& C\!-\!\!& C\!-\!H \\
\diagup & | & | & | \\
H & H & H & H
\end{array}
$$

is 3-bromopropan-1-ol.

The structural formula shows the arrangement of the atoms in a molecule in a simplified form, without showing the bonds. For example, the structural formula of propane is $CH_3CH_2CH_3$. If the molecule contains a side chain, the side chain is shown in brackets. So, the structural formula of

$$
\begin{array}{ccccc}
& H & H & Br & H \\
& | & | & | & | \\
H\!-\!\!& C\!-\!\!& C\!-\!\!& C\!-\!\!& C\!-\!H \\
& | & | & | & | \\
& H & H & & H \\
& & & H\!-\!C\!-\!H & \\
& & & | & \\
& & & H &
\end{array}
$$

is $CH_3CH_2CBr(CH_3)CH_3$, or $CH_3CH_2CBr(CH_3)_2$.

Remember that carbon always forms four covalent bonds.

Question 5.2

(i) Explain what is meant by the term *homologous series*.

(ii) State three characteristics of a homologous series.

(iii) Give the general formula of alkanes.

Answer

(i) A homologous series is a family of organic compounds with the same functional group but different carbon chain lengths.

(ii)   (1) Members of a homologous series have the same general formula.

(2) Members of a homologous series have similar chemical properties.

(3) There is a gradual change in the physical properties as the length of the carbon chain increases.

(iii) C_nH_{2n+2}

Background

A functional group is a group of atoms attached to a hydrocarbon chain and which determines the reactivity of the compound. For example, the C = O double bond is the functional group of carbonyl compounds, and —OH is the functional group of alcohols. A homologous series is a series of organic compounds with the same functional group. The characteristics of homologous series are:

1. Each member of a homologous series differs from the next by CH_2.
2. A general formula applies to all the members of a given homologous series. For example, alkenes have the general formula C_nH_{2n} and alcohols are $C_nH_{2n+2}O$ where n is an integer.
3. Members of a homologous series have similar chemical properties since they have a common functional group and since the length of the carbon chain has little effect on chemical reactivity.
4. The length of the carbon chain affects the physical properties of the members of a homologous series, such as melting and boiling points and solubility.

Question 5.3

(i) State what is meant by the term *isomers*.

(ii) Name and draw the structures of two branched structural isomers of pentane.

(iii) Does pentane have any other branched structural isomers? Explain your answer.

Answer

(i) Isomers are compounds with the same molecular formula but whose atoms are arranged differently.

(ii)

| 2-methylbutane | dimethylpropane |

(iii) Pentane has no other branched structural isomers because the atoms cannot be arranged differently from pentane and the two isomers shown above in (ii).

Background

If two organic molecules have the same molecular formula but differ in the arrangement of their atoms, they are called *isomers*.

Pentane has five carbon atoms arranged in an unbranched chain. The five carbon atoms can also be arranged in two branched arrangements: 2-methylbutane and 2,2-dimethylpropane (or dimethylpropane since the only location for a side chain here is on the second carbon atom).

To find the structural isomers of an alkane, try to work out all the possible ways of arranging the carbon atoms and name each resulting structure. You may find that some of the structures you obtain have the same name, which means that they are not two different isomers but the same molecule. Try five carbon atoms and you will find that the only ways they can be arranged are (i) in an unbranched chain (pentane), (ii) a chain of four carbon atoms with one methyl side chain (2-methylbutane), and (iii) a chain of three carbon atoms with two methyl side chains on the middle carbon (dimethylpropane).

Question 5.4

(i) State what is meant by the term *structural isomers*.

(ii) Name and draw the structures of three branched chain isomers of hexane.

(iii) Name the type of structural isomerism exhibited by the following pairs of isomers:

Isomer 1	Isomer 2	Type of isomerism

Answer

(i) Structural isomers are molecules with the same molecular formula but different structural formulae.

(ii)

| 2-methylpentane | 2,3-dimethylbutane | 2,2-dimethylbutane |

(iii)

Isomer 1	Isomer 2	Type of isomerism
		Functional group isomers
		Position isomers
		Chain isomers

Background

Structural isomers are molecules with the same molecular formula but different structures. There are three types of structural isomers:

Chain isomers occur when the hydrocarbon chain can either be straight or branched (see Question 5.3).

Position isomers occur when the functional group can be in different places.

Functional group isomers occur when two molecules have the same general formula but different functional groups.

Question 5.5

Alkenes are a group of unsaturated hydrocarbons with the general formula C_nH_{2n}.

      i.   **Explain what is meant by the terms *'hydrocarbon'* and *'unsaturated'*.**

      ii.   **Give the molecular formula of the alkene which contains 5 carbon atoms.**

      iii.   **Draw the structures of two positional isomers of this alkene.**

Answer

(i) A hydrocarbon is a compound containing only carbon and hydrogen. Unsaturated means that the compound contains at least one C = C double bond.

(ii) C_5H_{10}

(iii)

Background

The general formula of alkenes is C_nH_{2n}. the general formula of a homologous series allows us to find the molecular formula of any member of the series given the number of carbon atoms in the molecule. For example, the general formula of alkenes tells us that if the alkene contains 5 carbon atoms ($n = 5$), then the number of hydrogen atoms is $2n = 2\times5=10$. So the molecular formula is C_5H_{10}.

One possible structure of C_5H_{10} is

pent-1-ene (isomer 1)

We can also place the double bond between the second and third carbons, thus

H—C—C=C—C—C—H

pent-2-ene (isomer 2)

These two structures are positional isomers because they differ only in the position of the functional group (the C=C double bond). Note that these are the only positional isomers of C_5H_{10}. If we place the double bond between the third and fourth carbons, then we need to number the carbon atoms starting from the right, since the double bond is now closer to the right end of the molecule than to the left end:

H—C—C—C=C—C—H

pent-2-ene

This is the same as isomer 2 above.

Question 5.6

Intermolecular forces play a key role in determining the solubility of hydrocarbons in water. This is especially important in biological systems. The aqueous interior of a cell is kept from the surrounding, also aqueous, environment by a membrane which is constructed from molecules that have long hydrocarbon tails. Explain this in terms of intermolecular forces.

Answer

Hydrocarbons are insoluble in water because they are non-polar and cannot form effective bonds with water molecules. Therefore, the hydrocarbon tails of the molecules that make up the cell membrane cannot dissolve in water, thus keeping the aqueous interior of the cell separate from the cell's (exterior) aqueous environment.

Background

Polar molecules are soluble in polar solvents because they can be attracted to the solvent molecules by dipole forces of attractions. Non-polar molecules are insoluble in polar solvents and can only be attracted to similarly non-polar molecules.

The cell membrane is constructed from molecules which have a polar head and a long non-polar tail. The polar head is hydrophilic (water-loving), while the hydrocarbon tail is hydrophobic (water-hating). Therefore, the hydrophilic head is attracted to water molecules while the hydrophobic tail shuns water and seeks to aggregate with other hydrophobic molecules. The inside of the cell is an aqueous medium and so is the surrounding environment around the cell. Therefore, the molecules of the membrane are arranged in two layers in which the hydrophilic heads face the water from both sides of the membrane, while the hydrophobic tails lie next to one another in the interior of the membrane (like the filling in a sandwich).

This arrangement keeps the aqueous interior of the cell separate from the aqueous medium outside the cell.

6. ALKANES

Question 6.1

Alkanes containing 1-4 carbon atoms are gases at room temperature, while alkanes with 5-18 carbon atoms are liquids, and longer chain alkanes are solids. Explain this observation in terms of intermolecular forces.

Answer

As the length of the carbon chain increases, the number of electrons increases and so the strength of the van der Waals forces increases. This means that more energy is needed to overcome these forces, thus leading to higher melting and boiling points. Alkanes with up to 4 carbon atoms have weak van der Waals forces and so they are gases at room temperature. Alkanes with 5-18 carbons have stronger forces and are therefore liquids. Those with more than 18 carbons have van der Waals forces that are strong enough to hold the molecules together as a solid.

Background

Alkanes have weak van der Waals forces between their molecules. The larger the molecule, the more electrons it contains and the stronger the van der Waals forces. Therefore, the melting and boiling points increase as the length of the carbon chain increases.

Question 6.2

(i) Draw the structure of a branched chain isomer of hexane. Name this isomer.

(ii) Which of the two isomers in part (i) will have a higher melting point? Explain your answer.

Answer

3-methylpentane

(ii) Hexane has a higher melting point than 3-methylpentane. The branched chain molecules of 3-methylpentane cannot pack together as closely as the unbranched chain molecules of hexane, and so the van der Waals forces between the molecules of 3-methylpentane are weaker and therefore require less energy to break.

Background

Any branched chain isomer of hexane (6 carbon atoms) will have at least one alkyl group as a side chain. For example, we can have a chain of five carbon atoms with a methyl group attached to the second carbon, forming 2-methylpentane. Another isomer is 3-methylpentane. A chain of four carbon atoms with two methyl side groups attached to the second carbon makes yet another isomer, namely 2,2-dimethylbutane.

Branched chain molecules do not pack together as well as unbranched chain molecules, and therefore they have weaker van der Waals forces, which means lower melting and boiling points. For example, the boiling point of 3-methylpentane is about 22 degrees Celsius lower than that of unbranched hexane.

Question 6.3

Crude oil, a major source of fuel today, is composed mainly of alkanes. Explain briefly how crude oil was formed, and hence explain why it is described as a non-renewable source of energy.

Answer

Crude oil was formed by the decay of plant and animal remains at the high pressures and temperatures deep below the Earth's surface over millions of years. Crude oil is a non-renewable source of fuel because it is a finite source and cannot be replaced.

Background

Crude oil was formed in the Earth's crust over 300 million years ago from the remains of tiny marine organisms which sank to the bottom of the oceans and were covered by layers of rock. The increased temperature and pressure then caused these remains to decay, forming crude oil. There is a finite amount of oil in the Earth which cannot be replaced. Therefore, crude oil is a finite or non-renewable source of energy.

Question 6.4

Describe how fractional distillation is used to separate crude oil into useful fractions.

Answer

Fractional distillation depends on the fact that the different components of crude oil have different boiling points.

The crude oil is first vaporised by heating it in a furnace and is then passed into a fractionating tower. The temperature inside the tower decreases from the bottom towards the top. As the vapour rises upwards, it passes through a series of trays that get cooler towards the top. Different fractions of the vapour condense at different trays, depending on their boiling points. The liquids formed by condensation are then piped off separately. The shorter chain hydrocarbons condense nearer to the top of the tower since they have lower boiling points. Hydrocarbons with the highest boiling points are collected as a thick residue at the bottom of the tower.

Background

Fractional distillation is used to separate a mixture (gas or liquid) into its components. The separation is possible because the components have different boiling points. For example, suppose we are to separate a mixture of three substances: A, B and C. The boiling points of these three substances are 25, 100, and 200°C respectively. First, we vaporise the mixture. The resulting vapour is passed into a fractionating tower, which contains trays at different temperatures. The temperature of the trays decreases as we go up the tower. Suppose that the tray at which the vapour mixture is admitted into the tower is at a temperature of 250°C. The next (upper) tray is at 200°C, which is the boiling point of component C. Therefore, when the vapour mixture arrives at this tray, component C turns into a liquid and is separated from the vapour mixture. Components A and B remain in the vapour mixture and keep rising in the tower. When the mixture reaches a tray with a temperature of 100°C, component B condenses into a liquid and is separated from the vapour. The remaining vapour contains only A. We have thus separated the three components A, B and C.

Question 6.5

Petrol and kerosene are two fractions obtained by fractional distillation of crude oil. Petrol is a mixture of hydrocarbons with 5-10 carbon atoms, while kerosene is a mixture of hydrocarbons with 11-16 carbon atoms. Which of the two fractions is obtained nearer to the top of the fractionating tower? Explain your answer.

Answer

Petrol is obtained nearer to the top of the tower. As the length of the carbon chain increases, the van der Waals forces become stronger and therefore the boiling point increases. The hydrocarbons in petrol are shorter than those in kerosene, and so they have lower boiling points. Since the temperature inside a fractionating tower decreases from the bottom towards the top, the shorter chain molecules will condense nearer to the top of the tower.

Background

The temperature inside a fractionating tower decreases from the bottom to the top of the tower. The components of petrol have lower boiling points than those of kerosene, and so petrol will condense at a lower temperature compared to kerosene. Therefore, petrol is obtained near the top of the tower where the temperature is sufficiently low for the hydrocarbons that make up petrol to condense into a liquid fraction.

Question 6.6

(i) Define the process of cracking.

(ii) Explain the economic reasons for cracking.

(iii) Complete the following equation which represents the cracking of dodecane:

$$C_{12}H_{26} \longrightarrow C_8H_{18} + 2.....$$

(iv) Name the homologous series to which each of the two different products formed in the above reaction belongs. State one use for each of the two products.

Answer

(i) Cracking is the process by which long chain alkane molecules are broken into more useful, shorter chain molecules.

(ii) Crude oil has more of the longer chain hydrocarbons than is wanted, while the shorter chain hydrocarbons are in huge demand and are economically more valuable. So, cracking of the long chain alkanes produces shorter, more useful alkanes in order to meet the demand for such products, especially petrol. Cracking also produces alkenes which are used as chemical feedstock to make a huge range of useful products.

(iii) $C_{12}H_{26} \longrightarrow C_8H_{18} + 2C_2H_4$

(iv) C_8H_{18} is an alkane that is used in petrol. C_2H_4 is an alkene that is used to make plastics.

Background

Cracking is a very useful industrial process. In this process, long chain hydrocarbons are broken into smaller chain molecules that form more useful products.

One of the fractions produced by the fractional distillation of crude oil is the naphtha fraction (petrol), which is made up of relatively short chain alkanes. The amount of naphtha produced by distillation does not meet the huge demand for this product. On the other hand, distillation produces more than is wanted of the longer chain fractions. Therefore, it makes economic sense to convert the longer chain fractions into smaller chain products. This is achieved by cracking.

Cracking of a long chain alkane usually produces a shorter chain alkane as well as an alkene. For example, cracking of decane $C_{10}H_{22}$ can produce octane (a shorter chain alkane that is used in petrol) and ethene (an alkene used to make a wide range of everyday materials such as plastics):

$$C_{10}H_{22} \longrightarrow C_8H_{18} + C_2H_4$$

Cracking of an alkane can sometimes produce more than one molecule of the same alkene, or even two different alkenes. For example,

$$C_{12}H_{26} \longrightarrow C_8H_{18} + 2C_2H_4$$

Question 6.7

Name the two types of cracking, and state which of the two types is more likely to produce a high proportion of alkenes.

Answer

The two types of cracking are thermal cracking and catalytic cracking. Thermal cracking is more likely to produce a high proportion of alkenes.

Question 6.8

Outline the process of catalytic cracking. Your answer should include the conditions used and the major products of the process.

Answer

Catalytic cracking takes place at relatively low pressures and temperatures. A zeolite catalyst is used which consists of silicon dioxide and aluminium oxide. The catalyst has a honeycomb structure with a large surface area. The major products of catalytic cracking are branched alkanes, cycloalkanes, and aromatic compounds.

Question 6.9

Outline the process of thermal cracking. Your answer should include the conditions used, the type of reactive intermediates involved in the reaction, and the products formed.

Answer

Thermal cracking is carried out at high temperatures between 700-1200 K and high pressures up to 7000 kPa. The reactive intermediates formed in the reaction are free radicals. Thermal cracking produces alkanes and a high proportion of alkenes.

Question 6.10

Write an equation for the cracking of the alkane with 20 carbon atoms to produce ethene and propene in a 3:2 molar ratio together with one other product.

Answer

$$C_{20}H_{42} \rightarrow 3C_2H_4 + 2C_3H_6 + C_8H_{18}$$

Background

$C_{20}H_{42}$ can be cracked into three different products, two alkenes and one alkane. Ethene and propene are alkenes, so the third product must be an alkane. There are 20 carbon atoms in $C_{20}H_{42}$. Ethene and propene are produced in a 3:2 ratio (3 molecules of ethene and 2 molecules of propene). Three molecules of C_2H_4 contain a total of 6 carbon atoms, and two molecules of C_3H_6 also contain a total of 6 carbon atoms. This means that the third product (the alkane) must contain 8 carbon atoms, i.e. C_8H_{18}. The number of hydrogen atoms in this formula can be found either from the general formula of alkanes (C_nH_{2n+2}) or as the difference between the number of hydrogen atoms in $C_{20}H_{42}$ and the total number of hydrogen atoms in the alkenes formed.

Write a balanced equation for the complete combustion of ethane.

Answer

$$C_2H_6 + 3\frac{1}{2}O_2 \longrightarrow 2CO_2 + 3H_2O$$

Background

The complete combustion of any hydrocarbon produces carbon dioxide, water vapour, and a large amount of heat. This is why hydrocarbons are used as fuels. To balance a combustion reaction, always start by balancing C, then H, and finally O.

Question 6.12

When a Bunsen burner is used with a closed air hole, a black sooty deposit appears on the burner. Explain this in terms of the combustion reaction taking place in the burner.

Answer

Due to the limited supply of oxygen when the air hole is closed, the fuel in the burner undergoes incomplete combustion which produces soot.

Background

Complete combustion requires an excess supply of oxygen. However, when the amount of oxygen is limited, the combustion is described as incomplete and carbon monoxide is produced

$$C_2H_6 + 2\frac{1}{2}O_2 \longrightarrow 2CO + 3H_2O$$

With even less oxygen, incomplete combustion produces carbon (soot):

$$C_2H_6 + 1\frac{1}{2}O_2 \longrightarrow 2C + 3H_2O$$

The burning of hydrocarbon-based fuels produces polluting products. For each of the following pollutants, explain how the pollutant is released and state one adverse effect it has on the environment or on health.

CO_2, CO, NO_x, SO_2, H_2O

Answer

CO_2: a product of the complete combustion of fuel. It's a greenhouse gas that contributes to global warming.

CO: produced by the incomplete combustion of fuel. CO is poisonous.

NO_x: produced when nitrogen and oxygen from the air react together at the high temperatures produced by the car engine. Nitrogen oxides react with water and oxygen to form acid rain which can cause damage to buildings.

SO_2: produced from the sulphur-containing impurities present in the fuel. SO_2 reacts with water and oxygen to form acid rain.

H_2O: is produced by the combustion of fuel. It is a greenhouse gas that contributes to global warming.

One of the pollutants produced by a petrol engine is nitrogen monoxide, which contributes to acid rain. As petrol itself does not contain nitrogen, how do you explain the formation of nitrogen monoxide in a petrol engine? Your answer should include an equation for the formation of nitrogen monoxide.

Answer

The combustion of petrol produces high temperatures in the engine. At high temperatures, nitrogen and oxygen in the air react forming nitrogen monoxide:

$$N_2 + O_2 \longrightarrow 2NO$$

Background

Although petrol itself does not contain nitrogen, the combustion of petrol in a car engine produces enough heat for nitrogen and oxygen in the air to react, thus producing nitrogen monoxide. Nitrogen monoxide can then react with oxygen and water vapour in the air to form nitric acid, causing acid rain.

Question 6.15

Sulphur dioxide SO_2 is a gas that contributes to acid rain. It is produced by the burning of sulphur-containing fuels.

(i) Write an equation to show the formation of acid rain from SO_2.

(ii) Explain how SO_2 can be removed from the exhaust gas of power stations. State a use for the substance formed in the process of removing SO_2.

Answer

(i) $SO_2 + H_2O + \dfrac{1}{2}O_2 \longrightarrow H_2SO_4$

(ii) Calcium oxide, CaO, is used to absorb sulphur dioxide. This produces gypsum, $CaSO_4$, which is used as plaster.

Question 6.16

All new cars with petrol engines are now fitted with catalytic converters.

(i)   Name the metal catalysts used in the catalytic converter.
(ii) Explain why the catalytic converter has a honeycomb structure.
(iii) Write a balanced equation for the reaction responsible for removing both carbon monoxide and nitrogen monoxide in a catalytic converter.
(iv) Write a balanced equation for the reaction in which unburnt octane C_8H_{18} reacts with nitrogen monoxide producing nitrogen, carbon dioxide, and water vapour.

Answer

(i) Platinum and rhodium.

(ii) The honeycomb shape provides an enormous surface area for the reactions to take place, so a little of the expensive catalyst goes a long way.

(iii) $2CO + 2NO \longrightarrow N_2 + 2CO_2$

(iv) $C_8H_{18} + 25NO \longrightarrow 12\frac{1}{2}N_2 + 8CO_2 + 9H_2O$

Background

The catalytic converter is used to convert carbon monoxide, nitrogen oxides, and unburnt hydrocarbons into less harmful products. The converter uses platinum and rhodium metals as catalysts. Since the catalyst is expensive, the converter is made of a ceramic material coated with the catalyst. The converter has a honeycomb shape in order to maximise the surface area of the catalyst. The larger the surface area of the catalyst, the faster the reactions.

Question 6.17

Carbon dioxide is an important greenhouse gas, without which the Earth would be too cold to sustain life. However, increased levels of this gas in the atmosphere are believed to be the major cause of global warming. State the main source of this increase in carbon dioxide levels, and briefly explain how such increased levels can lead to global warming.

Answer

Increased levels of carbon dioxide in the atmosphere are caused by burning of fossil fuels to produce energy. Carbon dioxide traps infra-red radiation from the sun so that the Earth's surface heats up, thus causing global warming.

Background

Global warming is an observed increase in the temperature of the atmosphere believed to be caused by the build-up of greenhouse gases. Greenhouse gases are gases which trap heat in the atmosphere.

When the sun's radiation is reflected off the surface of the Earth towards the outer space, part of this radiation is absorbed by certain gases in the atmosphere. This radiation then becomes 'trapped' in the atmosphere, leading to an increase in temperature. This greenhouse effect is critical because it helps maintain a relatively warm environment for life on Earth. However, increased levels of greenhouse gases may lead to too much warming that could have disastrous consequences.

www.ingramcontent.com/pod-product-compliance
Lightning Source LLC
Chambersburg PA
CBHW061514180526
45171CB00001B/173